Advanced Mathematical Economics

T0186226

As the intersection between economics and mathematics continues to grow in both theory and practice, a solid grounding in mathematical concepts is essential for all serious students of economic theory.

In this clear and entertaining volume, Rakesh V. Vohra sets out the basic concepts of mathematics as they relate to economics. The book divides the mathematical problems that arise in economic theory into three types: feasibility problems, optimality problems and fixed-point problems. Of particular salience to modern economic thought are sections on lattices, supermodularity, matroids and their applications. In a departure from the prevailing fashion, much greater attention is devoted to linear programming and its applications.

Of interest to advanced students of economics as well as those seeking a greater understanding of the influence of mathematics on 'the dismal science'. *Advanced Mathematical Economics* follows a long and celebrated tradition of the application of mathematical concepts to the social and physical sciences.

Rakesh V. Vohra is the John L. and Helen Kellogg Professor of Managerial Economics and Decision Sciences at the Kellogg School of Management at Northwestern University, Illinois.

Routledge advanced texts in economics and finance

Advanced Mathematical Economics

Rakesh V. Vohra

Routledge
Taylor & Francis Group

LONDON AND NEW YORK

First published 2005
by Routledge
2 Park Square, Milton Park, Abingdon, Oxon, OX14 4RN

Simultaneously published in the USA and Canada
by Routledge
270 Madison Ave, New York NY 10016

Routledge is an imprint of the Taylor & Francis Group

Transferred to Digital Printing 2010

© 2005 Rakesh V. Vohra

Typeset in Times New Roman by
Newgen Imaging Systems (P) Ltd, Chennai, India

British Library Cataloguing in Publication Data
A catalogue record for this book is available
from the British Library

Library of Congress Cataloging in Publication Data
A catalog record for this book has been requested

ISBN 0–415–70007–8 (hbk)
ISBN 0–415–70008–6 (pbk)

Contents

Preface

I wanted to title this book 'Leisure of the Theory Class'. The publishers demurred. My second choice was 'Feasibility, Optimality and Fixed Points'. While accurate, it did not identify, as the publisher noted, the intended audience. We settled at last on the anodyne title that now graces this book. As it suggests, the book is about mathematics. The qualifier 'advanced' signifies that the reader should have some mathematical sophistication. This means linear algebra and basic real analysis.[1] Chapter 1 provides a list of cheerful facts from these subjects that the reader is expected to know. The last word in the title indicates that it is directed to students of the *dismal science*.[2]

Three kinds of mathematical questions are discussed. Given a function f and a set S,

- Find an x such that $f(x)$ is in S. This is the **feasibility** question.
- Find an x in S that optimizes $f(x)$. This is the problem of **optimality**.
- Find an x in S such that $f(x) = x$; this is the **fixed point** problem.

These questions arise frequently in Economic Theory, and the applications described in the book illustrate this.

The topics covered are standard. Exceptions are matroids and lattices. Unusual for a book such as this is the attention paid to Linear Programming. It is common amongst cyclopean Economists to dismiss this as a special case of Kuhn–Tucker. A mistake in my view. I hope to persuade the reader, by example, of the same. Another unusual feature, I think, are the applications. They are not merely computational, i.e., this is how one uses Theorem X to compute such and such. They are substantive pieces of economic theory. The homework problems, as my students will attest, are not for the faint hearted.

Of the making of books there is no end.[3] The remark is particularly true for books devoted to topics that the present one covers. So, some explanation is required of how this book differs from others of its ilk.

The voluminous Simon and Blume (1994), ends (with exceptions) where this book begins. In fact a knowledge of Simon and Blume is a good prerequisite for this one.

The thick, square, Mas-Collel *et al.* (1995) contains an appendix that covers a subset of what is covered here. The treatment is necessarily brief, omitting many interesting details and connections between topics.

Sundaram's excellent 'First Course in Optimization' (1996) is perhaps closest of the more recent books. But there are clear differences. Sundaram covers dynamic optimization while this book does not. On the other hand, this book discusses fixed points and matroids, while Sundaram does not.

This book is closest in spirit to books of an earlier time when, giants, I am reliably informed, walked the Earth. Two in particular have inspired me. The first is Joel Franklin's 'Methods of Mathematical Economics' (1980), a title that pays homage to Courant and Hilbert's celebrated 'Mathematical Methods of Physics' (1924). Franklin is a little short on the economic applications of the mathematics described. However, the informal and direct style convey his delight in the subject. It is a delight I share and I hope this book will infect the reader with the same.

The second is Nicholas Rau's 'Matrices and Mathematical Programming: An Introduction for Economists'. Rau is formal, precise and startlingly clear. I have tried to match this standard, going so far as to follow, in some cases, his exposition of proofs.

The book before you is the outgrowth of a PhD class that all graduate students in my department must take in their first year.[4] Its roots however go back to my salad days. I have learnt an enormous amount from teachers and colleagues much of which infuses this book. In no particular order, they are Ailsa Land, Saul Gass, Bruce Golden, Jack Edmonds, H. Peyton Young, Dean Foster, Teo Chung Piaw and James Schummer.

Four cohorts of graduate students at Kellogg and other parts of Northwestern have patiently endured early versions of this book. Their questions, both puzzled and pointed have been a constant goad. I hope this book will do them justice.

Denizens of MEDS, my department, have patiently explained to me the finer points of auctions, general equilibrium, mechanism design and integrability. In return I have subjected them to endless speeches about the utility of Linear Programming. The book is as much a reflection of my own hobby horses as the spirit of the department.

This book could not have been written without help from both my parents (Faqir and Sudesh Vohra) and in-laws (Harish and Krishna Mahajan). They took it in turns to see the children off to school, made sure that the fridge was full, dinner on the table and the laundry done. If it continues I'm in clover! My wife, Sangeeta took care of many things I *should* have taken care of, including myself.

Piotr Kuszewski, a graduate student in Economics, played an important role in preparing the figures, formatting and producing a text that was a pleasure to look at.

Finally, this book is dedicated to my children Akhil and Sonya, who I hope will find the same joy in Mathematics as I have. Perhaps, like the pilgrims in

Flecker's poem, they will go

> Always a little further: it may be
> Beyond the last blue mountain barred with snow,
> Across that angry or that glimmering sea,
> White on a throne or guarded in a cave
> There lives a prophet who can understand
> Why men were born: but surely we are brave,
> Who take the Golden Road to Samarkand.

Notes

1 An aside, this book has many. The writer Robert Heinlein, suggested that the ability to solve a quadratic be a minimal condition to be accorded the right to vote:

> ... step into the polling booth and find that the computer has generated a new quadratic equation just for you. Solve it, the computer unlocks the voting machine, you vote. But get a wrong answer and the voting machine fails to unlock, a loud bell sounds, a red light goes on over that booth – and you slink out, face red, you having just proved yourself too stupid and/or ignorant to take part in the decisions of the grownups. Better luck next election! No lower age limit in this system – smart 12-year-old girls vote every election while some of their mothers – and fathers – decline to be humiliated twice.

2 The term is due to Thomas Carlyle. Oddly, his more scathing description of Economics, a *pig philosophy*, has never caught on. Other nineteenth century scribblers like John Ruskin called economics the *bastard science* while Matthew Arnold referred to economists as a *one eyed race*.

3 The line is from *Ecclesiastes* 12:12. It continues with 'and much reading is a weariness of the soul'.

4 One can usefully cover the first seven chapters (if one is sparing in the applications) in a 1 quarter class (10 weeks, 3.5 h a week). The entire book could be exhausted in a semester.

References

Courant, R. and Hilbert, D.: 1924, *Methoden der mathematischen Physik*, Springer, Berlin.

Franklin, J. N.: 1980, *Methods of mathematical economics: linear and nonlinear programming: fixed-point theorems*, Undergraduate texts in mathematics, Springer-Verlag, New York.

Heinlein, R. A.: 2003, *Expanded universe*, Baen; Distributed by Simon & Schuster, Riverdale, NY, New York.

Mas-Colell, A., Whinston, M. D. and Green, J. R.: 1995, *Microeconomic theory*, Oxford University Press, New York.

Rau, N.: 1981, *Matrices and mathematical programming: an introduction for economists*, St. Martin's Press, New York, N.Y.

Simon, C. P. and Blume, L.: 1994, *Mathematics for economists*, 1st edn, Norton, New York.

Sundaram, R. K.: 1996, *A first course in optimization theory*, Cambridge University Press, Cambridge, New York.

1   Things to know

This chapter summarizes notation and mathematical facts used in the rest of the book. The most important of these is 'iff' which means if and only if.

1.1   Sets

Sets of objects will usually be denoted by capital letters, A, S, T for example, while their members by lower case letters (English or Greek). The empty set is denoted $\varnothing$. If an object x belongs to a set S we write $x \in S$ and if it does not we write $x \notin S$. The set of objects not in a given set S is called the **complement** of S and denoted S^c. Frequently our sets will be described by some property shared by all its elements. We will write this in the form $\{x: x$ has property P$\}$.

The elements in common between two sets, S and T, their **intersection**, is denoted $S \cap T$. Elements belonging to one or the other or both sets, their **union**, is denoted $S \cup T$. The set of elements belonging to S but not T is denoted $S \setminus T$. If the elements of a set S are entirely contained in another set T, we say that S is a **subset** of T and write $S \subseteq T$. If S is strictly contained in T, meaning there is at least one element of T not in S we write $S \subset T$. In this case we say that S is a **proper** subset of T. The number of elements in a set S, its **cardinality**, is denoted $|S|$.

The upside down 'A', $\forall$, means 'for all' while the backward 'E', $\exists$, means 'there exists'.

1.2   The space we work in

This entire book is confined to the space of vectors of real numbers (also called points) with n components. This space is denoted $\mathbb{R}^n$. The non-negative orthant, the set of vectors all of whose components are non-negative, is denoted $\mathbb{R}^n_+$. The jth component of a vector x will be denoted x_j, while the jth vector from some set will be denoted x^j. If x and y in $\mathbb{R}^n$ then:

- $x = y$ iff $x_i = y_i$ for all i,
- $x \geq y$ iff $x_i \geq y_i$ for all i,
- $x > y$ iff $x_i \geq y_i$ for all i with strict inequality for at least one component, and
- $x \gg y$ iff $x_i > y_i$ for all i.

1.3   Facts from real analysis

Definition 1.1   *Given a subset S of real numbers, the **supremum** of S, written* $\sup(S)$, *is the smallest number that is larger than every number in S. The **infimum** of S, written* $\inf(S)$, *is the biggest number that is smaller than every number in S.*

For example, if $S = \{x \in \mathbb{R}^1 : 0 < x < 1\}$, then $\sup(S) = 1$ and $\inf(S) = 0$. Notice that neither the infimum or supremum of S are contained in S.

If x and y are any two vectors in $\mathbb{R}^n$, we will denote by $d(x, y)$ the Euclidean distance between them, i.e.,

$$d(x, y) = \sqrt{\sum_{j=1}^{n}(x_j - y_j)^2}.$$

The length of the vector x is just $d(x, 0)$ and is sometimes written $\|x\|$. A **unit vector** x is one whose length is 1, i.e., $\|x\| = 1$. The **dot product** of two vectors x and y is denoted $x \cdot y$ or xy and is defined thus:

$$x \cdot y = \sum_{j=1}^{n} x_j y_j = d(x, 0)\, d(y, 0) \cos\theta,$$

where θ is the angle between x and y. Notice that $d(x, 0)^2 = x \cdot x$. A pair of vectors x and y is called **orthogonal** if $x \cdot y = 0$.

Definition 1.2   *The sequence* $\{x^k\}_{k \geq 1} \in \mathbb{R}^n$ **converges** *to* $x^0 \in \mathbb{R}^n$ *if for every* $\epsilon > 0$ *there is an integer K (possibly depending on* ϵ) *such that*

$$d(x^k, x^0) < \epsilon, \quad \forall k \geq K.$$

If $\{x^k\}_{k \geq 1}$ *converges to* x^0 *we write* $\lim_{k \to \infty} x^k = x^0$.

Example 1   *Consider the following sequence of real numbers:* $x^k = 1/k$. *The limit of this sequence is 0. To see why, fix an* $\epsilon > 0$. *Now, can we choose a k large enough so that* $|1/k - 0| < \epsilon$? *In this case, yes. Simply pick any* $k > 1/\epsilon$.

There are a host of tricks and techniques for establishing when a sequence has a limit and what that limit is. We mention one, called the **Cauchy criterion**.[1]

Theorem 1.3   *Let* $\{x^m\}$ *be a sequence of vectors in* $\mathbb{R}^n$. *Suppose for any* $\epsilon > 0$ *there is a N sufficiently large such that for all* $p, q > N$, $d(x^p, x^q) < \epsilon$. *Then* $\{x^m\}$ *has a limit.*

It will often be the case that we will be interested in a sequence $\{x^k\}_{k \geq 1}$ all of whose members are in some set S and will want to know if its limit (should it exist) is in S. As an example, suppose S is the set of real numbers *strictly* between 0 and 1. Consider the sequence $x^k = 1/(k + 1)$, every element of which is in S. The limit of this sequence is 0, which is not in S.

Definition 1.4 *A set $S \subset \mathbb{R}^n$ is said to be **closed** if it contains all its limit points. That is, if $\{x^k\}_{k \geq 1}$ is any convergent sequence of points in S, then $\lim_{k \to \infty} x^k$ is in S as well.*

Example 2 *We prove that the set $\{x \in \mathbb{R}^1: 0 \leq x \leq 1\} = [0, 1]$ is closed. Let $\{x^k\}_{k \geq 1} \in [0, 1]$ be a convergent subsequence with limit x^0. Suppose for a contradiction that $x^0 \notin [0, 1]$. In fact we may suppose that $x^0 > 1$. Pick $\epsilon = (x^0 - 1)/2 > 0$. Since $\lim_{k \to \infty} x^k = x^0$, for any $\epsilon > 0$ there is a k sufficiently large such that $|x^k - x^0| \leq \epsilon$. For our choice of ϵ this implies that $x^k > 1$, a contradiction.*

Definition 1.5 *A set $S \subset \mathbb{R}^n$ is called **open** if for every $x \in S$ there is an $\epsilon > 0$ such that any y within distance of ϵ of x, $d(x, y) < \epsilon$, is in S.*

An important class of open and closed sets are called **intervals**. Given two numbers $a < b$, the closed interval $[a, b]$ is the set $\{x \in \mathbb{R}^1: a \leq x \leq b\}$. The open interval (a, b) is the set $\{x \in \mathbb{R}: a < x < b\}$.

A set can be neither open or closed, for example, $S = \{x \in \mathbb{R}^1: 0 < x \leq 1\}$. The sequence $\{1/k\}_{k \geq 1}$ has a limit that is not in this set. So, S is not closed. However, there is no $\epsilon > 0$ sufficiently small such that every point within distance of 1 is in S. Thus, S is not open.

A point $x \in S$ is called an **interior** point of S if the set $\{y: d(y, x) < \epsilon\}$ is contained in S for all $\epsilon > 0$ sufficiently small. It is called a **boundary** point if $\{y: d(y, x) < \epsilon\} \cap S^c$ is non-empty for all $\epsilon > 0$ sufficiently small. The set of all boundary points of S is called the boundary of S.

Example 3 *Consider the set $S = \{x \in \mathbb{R}^n: d(x, 0) \leq r\}$. Its interior is $\{x: d(x, 0) < r\}$ while its boundary is $\{x: d(x, 0) = r\}$.*

Here is a list of important facts about open and closed sets:

1. a set $S \subset \mathbb{R}^n$ is open if and only if its complement is closed;
2. the union of any number of open sets is open;
3. the intersection of a *finite* number of open sets is open;
4. the intersection of a any number of closed sets is closed;
5. the union of a *finite* number of closed sets is closed.

If $S \subset \mathbb{R}^1$ is closed, the infimum and supremum of S are in S. In fact, they coincide with the smallest and largest member of S, respectively.

Definition 1.6 *The **closure** of a set S is the set S combined with all points that are the limits of sequences of points in S.*

Definition 1.7   *A set $S \subset \mathbb{R}^n$ is called* **bounded** *if there is a finite positive number r such that $\|x\| \leq r$ for all $x \in S$. It is called* **compact** *if it is both closed and bounded.*

Theorem 1.8 (Bolzano–Weierstrass)   *Let S be a bounded set and $\{x_n\}$ an infinite sequence all of whose elements lie in S. Then the sequence $\{x_n\}$ contains a convergent subsequence.*

A real valued function f on $\mathbb{R}^n$ is a rule that assigns to each $x \in \mathbb{R}^n$ a real number. We denote this as $f: \mathbb{R}^n \to \mathbb{R}$. If we write $f: \mathbb{R}^n \to \mathbb{R}^m$ it means the function assigns to each element of $\mathbb{R}^n$ an element of $\mathbb{R}^m$.

Definition 1.9   *A real valued function f on $\mathbb{R}^n$ is* **continuous at the point** a *if for any $\epsilon > 0$ we can find a $\delta > 0$ (possibly depending on ϵ) such that for all x within distance δ of a, $|f(x) - f(a)| < \epsilon$. This is sometimes abbreviated as $\lim_{x \to a} f(x) = f(a)$.*

Definition 1.10   *A function is said to be* **continuous** *on the set $S \subset \mathbb{R}^n$ if for every $a \in S$ and any $\epsilon > 0$ we can find a $\delta > 0$ (possibly depending on ϵ and a) such that for all $x \in S$ within distance δ of a, $|f(x) - f(a)| < \epsilon$. The main point is that in $\lim_{x \to a} f(x)$ we require the sequence of points that converge to a to be entirely in S.*

Example 4   *We show that the function $f(x) = x^2$ where $x \in \mathbb{R}^1$ is continuous. Choose an $\epsilon > 0$ that is small and any $a \in \mathbb{R}^1$. Set $\delta = \epsilon/|3a|$ and notice that for any x within distance δ of a (i.e. $|x - a| \leq \delta$) we have that $|f(x) - f(a)| = |x^2 - a^2| = |(x - a)(x + a)| < \delta|x + a| \leq |3a|\delta \leq \epsilon$.*

You should be able to verify the following facts about continuous functions:

1.   the sum of two continuous functions is a continuous function;
2.   the product of two continuous functions is continuous;
3.   the quotient of two continuous functions is continuous at any point where the denominator is not zero.

The following lemma illustrates how the notion of open set and continuous function are related to each other.

Lemma 1.11   *Let $S \subset \mathbb{R}^n$ and $f: S \to \mathbb{R}$ be continuous on S. Let $K \subset \mathbb{R}$ be an open set and suppose $f^{-1}(K) = \{x \in S: f(x) \in K\} \neq \varnothing$. Then $f^{-1}(K)$ is an open set.*

Proof   Pick an $a \in f^{-1}(K)$. Since f is continuous, for all $\epsilon > 0$ sufficiently small there is a $\delta > 0$ such that for all x within distance δ of a, $|f(x) - f(a)| < \epsilon$. Since K is open it follows that $f(x) \in K$, implying that $x \in f^{-1}(K)$ proving the result.   ∎

Let $S \subset \mathbb{R}^n$ and M an index set. A collection of open sets $\{K_i : i \in M\}$ is an **open covering** of S if $S \subseteq \cup_{i \in M} K_i$.

Theorem 1.12 (Heine–Borel Theorem)   *A set $S \subset \mathbb{R}^n$ is compact iff every open covering of S contains a finite open subcovering.*

At various places in the book we will be interested in identifying an $x \in S$ that maximizes some real valued function f defined on S. We will write this problem as $\max_{x \in S} f(x)$. The set of possible points in S that solve this problem will be denoted $\arg\max_{x \in S} f(x)$. Similarly, $\arg\min_{x \in S} f(x)$ is the set of points in S that minimize $f(x)$. The next theorem provides a sufficient condition for the non-emptiness of $\arg\max_{x \in S} f(x)$ as well as $\arg\min_{x \in S} f(x)$.

Theorem 1.13 (Weierstrass maximum Theorem)   *Let $S \subset \mathbb{R}^n$ be compact and f a continuous real valued function on S. Then $\arg\max_{x \in S} f(x)$ and $\arg\min_{x \in S} f(x)$ exist and are both in S.*

Proof   Suppose the set $T = \{y \in \mathbb{R}^1 : \exists x \in S \text{ s.t. } y = f(x)\}$ is compact. Then $\sup(T) = \sup_{x \in S} f(x)$ and $\inf(T) = \inf_{x \in S} f(x)$. Since T is closed it follows that $\sup(T)$ and $\inf(T)$ are contained in T, i.e. there is an $x' \in S$ such that $f(x') = \sup_{x \in S} f(x)$ and an $x'' \in S$ such that $f(x'') = \inf_{s \in S} f(x)$. We now prove that T is compact.

Let M be an index set and $\{K_i\}_{i \in M}$ a collection of open sets that cover T. Since f is continuous, the sets $f^{-1}(K_i) = \{x \in S : \text{s.t.} f(x) \in K_i\}$ are open. Furthermore, by Lemma 1.11, the collection $\{f^{-1}(K_i)\}_{i \in M}$ forms an open cover of S. Compactness of S allows us to invoke the Heine–Borel theorem to conclude the existence of a finite subcover, with index set M' say. If $T \subseteq \cup_{i \in M'} K_i$ we are done. Consider any $y^* \in T$. Let x^* be such that $f(x^*) = y^*$. Notice that there is a $j \in M'$ such that $x^* \in f^{-1}(K_j)$. This implies that $y^* \in K_j$, i.e. $y^* \in \cup_{i \in M'} K_i$.   ∎

Definition 1.14   *The function $f : \mathbb{R} \to \mathbb{R}$ is **differentiable** at the point a if $(f(x) - f(a))/(x - a)$ has a limit as $x \to a$. The **derivative** of f at a is this limit and denoted $f'(a)$ or $\frac{df}{dx}\big|_{x=a}$.*

Every differentiable function is continuous, but the converse is not true.

If $f : \mathbb{R}^n \to \mathbb{R}$, the **partial derivative** of f with respect to x_j (when it exists) is the derivative of f with respect to x_j holding all other variables fixed. It is denoted $\frac{\partial f}{\partial x_j}$. The vector of partial derivatives, one for each component of x is called the **gradient** of f and denoted $\nabla f(x)$.

Theorem 1.15 (Rolle's Theorem)   *Let $f : \mathbb{R} \to \mathbb{R}$ be differentiable. For any $x < y$ there is a θ strictly between x and y such that*

$$f'(\theta) = \frac{f(y) - f(x)}{y - x}.$$

1.4   Facts from linear algebra

Given a set $S = \{x^1, x^2, \ldots\}$ of vectors, we will, in an abuse of notation use S to denote both the set of vectors as well as the index set of the vectors.

Definition 1.16   *A vector y can be expressed as a **linear combination** of vectors in $S = \{x^1, x^2, \ldots\}$ if there are real numbers $\{\lambda_j\}_{j \in S}$ such that*

$$y = \sum_{j \in S} \lambda_j x^j.$$

*The set of all vectors that can be expressed as a linear combination of vectors in S is called the **span** of S and denoted $\mathrm{span}(S)$.*

Definition 1.17   *A finite set $S = \{x^1, x^2, x^3, \ldots\}$ of vectors is said to be **linearly independent** (LI) if for all sets of real numbers $\{\lambda_j\}_{j \in S}$*

$$\sum_{j \in S} \lambda_j x^j = 0 \;\Rightarrow\; \lambda_j = 0, \quad \forall j \in S.$$

The following are examples of LI sets of vectors:

$$S = \{(1, -2)\},$$
$$S = \{(0, 1, 0), (-2, 2, 0)\},$$
$$S = \{(1, 1), (0, -3)\}.$$

A finite set S of vectors is said to be **linearly dependent** (LD) if it is not LI. This implies that there exist real numbers $\{\lambda_j\}_{j \in S}$ not all zero such that

$$\sum_{j \in S} \lambda_j x^j = 0.$$

Equivalently, one of the vectors in S can be expressed as a linear combination of the others. The following are examples of LD sets of vectors:

$$S = \{(1, -2), (2, -4)\},$$
$$S = \{(0, 1, 0), (-2, 2, 0), (-2, 3, 0)\},$$
$$S = \{(1, 1, 0), (0, -3, 1), (2, 5, -1)\}.$$

Definition 1.18   *The **rank** of a (not necessarily finite) set S of vectors is the size of the largest subset of linearly independent vectors in S.*

Ranks of the various sets of vectors above are listed below:

$S = \{(1, -2)\}$, rank $= 1$;

$S = \{(0, 1, 0), (-2, 2, 0)\}$, rank $= 2$;

$S = \{(1, 1), (0, -3)\}$, rank $= 2$;

$S = \{(1, -2), (2, -4)\}$, rank $= 1$;

$S = \{(0, 1, 0), (-2, 2, 0), (-2, 3, 0)\}$, rank $= 2$;

$S = \{(1, 1, 0), (0, -3, 1), (2, 4, -1)\}$, rank $= 3$.

Definition 1.19 *Let S be a set of vectors and $B \subset S$ be finite and LI. The set B of vectors is said to be a **maximal LI** set if the set $B \cup \{x\}$ is LD for all vectors $x \in S \setminus B$. A maximal LI subset of S is called a **basis** of S.*

Theorem 1.20 *Every $S \subset \mathbb{R}^n$ has a basis. If B is a basis for S, then $\text{span}(S) = \text{span}(B)$.*

Theorem 1.21 *Let $S \subset \mathbb{R}^n$. If B and B' are two bases of S, then $|B| = |B'|$.*

From this theorem we see that if S has a basis B, then the rank of S and $|B|$ coincide.

Definition 1.22 *Let S be a set of vectors. The **dimension** of $\text{span}(S)$ is the rank of S.*

The span of $\{(1, 0), (0, 1)\}$ is $\mathbb{R}^2$ and so the dimension of $\mathbb{R}^2$ is two. Generalizing this, we deduce that the dimension of $\mathbb{R}^n$ is n.

A rectangular array of numbers consisting of m rows and n columns is called an **m × n matrix**. It is usually denoted A and the entry in the ith row and jth column will be denoted a_{ij}. Whenever we use lower case Latin letters to denote the numbers appearing in the matrix, we use the corresponding upper case letters to denote the matrix. The ith row will be denoted a_i and the jth column will be denoted a^j. It will be useful later on to think of the columns and rows of A as vectors.

If it is necessary to emphasize the dimensions of a matrix A, we will write $A_{m \times n}$. If A is a matrix, its **transpose**, written A^T is the matrix obtained by interchanging the columns of A with its rows. The $n \times n$ matrix A where $a_{ij} = 0$ for all $i \neq j$ and $a_{ii} = 1$ for all i is called the **identity matrix** and denoted I.

The product of an $m \times n$ matrix A with a $n \times 1$ vector x, Ax is the $m \times 1$ vector whose ith component is $\sum_{j=1}^{n} a_{ij} x_j$. The ith component can also be written using dot product notation as $a_i \cdot x$. Similarly the jth component of yA will be $y \cdot a^j$ or $\sum_{i=1}^{m} a_{ij} y_i$.

The **inverse** of an $n \times n$ matrix A, is the matrix B such that $BA = I = AB$. The inverse of A is usually written A^{-1}. Not all matrices have inverses, and when they do, they are called **invertible**.

Associated with every $n \times n$ matrix A is another matrix called its **adjoint**, denoted $\mathrm{adj}(A)$. The reader may consult any standard text on linear algebra for a definition. Associated with every $n \times n$ matrix A is a real valued function called its **determinant** and denoted $|A|$. Again, the reader should consult a standard text for a definition. The inverse of a matrix A (when it exists) is related to its adjoint as follows:

$$A^{-1} = \frac{\mathrm{adj}(A)}{|A|}.$$

This relation is known as **Cramer's rule**.

In the sequel, we will be interested in the span of the columns (or rows) of a matrix. If A is a matrix we will write span(A) to denote the span of the columns of A and span(A^T) the span of the rows of A.

Definition 1.23   *The **kernel** or **null space** of A is the set $\{x \in \mathbb{R}^n \colon Ax = 0\}$.*

The following theorem summarizes the relationship between the span of A and its kernel.

Theorem 1.24   *If A is an $m \times n$ matrix then the dimension of the of span(A) plus the dimension of the kernel of A is n.*

This is sometimes written as

$$\dim[\mathrm{span}(A)] + \dim[\ker(A)] = n.$$

Since the dimension of the span of A and the rank of A coincide we can rewrite this as:

$$\mathrm{rank}(A) + \dim[\ker(A)] = n.$$

A similar expression holds for A^T:

$$\mathrm{rank}(A^T) + \dim[\ker(A^T)] = m.$$

The **column rank** of a matrix is the dimension of the span of its columns. Similarly, the **row rank** is the dimension of the span of its rows.

Theorem 1.25   *Let A be an $m \times n$ matrix. Then the column rank of A and A^T are the same.*

Thus the column and row rank of A are equal. This allows us to define the **rank** of a matrix A to be the dimension of span(A).

1.5   Facts from graph theory

A **graph** is a collection of two objects. The first is a finite set $V = \{1, \ldots, n\}$ called **vertices**. The second is a set E of (unordered) pairs of vertices called **edges**. As an example, suppose $V = \{1, 2, 3, 4\}$ and $E = \{(1, 2), (2, 3), (3, 4), (2, 4)\}$. A pictorial representation of this graph is shown in Figure 1.1. A graph is called a **complete graph** if E consists of every pair of vertices in V.

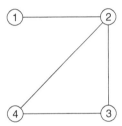

Figure 1.1

The **end points** of an edge $e \in E$ are the two vertices i and j that define that edge. In this case we write $e = (i, j)$. The **degree** of a vertex is the number of edges that contain it. In the graph above, the degree of vertex 3 is 2 while the degree of vertex 2 is 3. A pair $i, j \in V$ is called **adjacent** if $(i, j) \in E$.

Lemma 1.26   *The number of vertices of odd degree in a graph is even.*

Proof   Let O be the set of odd degree vertices in a graph and P the set of even degree vertices. Let d_i be the degree of vertex $i \in V$. If we add the degrees of all vertices we count all edges twice (because each edge has two endpoints), so

$$\sum_{i \in V} d_i = 2|E|.$$

Hence the sum of degrees is even. Now

$$\sum_{i \in V} d_i = \sum_{i \in O} d_i + \sum_{i \in P} d_i.$$

The second term on the right is an even number, while the first term is the sum of odd numbers. Since their sum is even, it follows that $|O|$ is an even number.   ■

Fix a graph $G = (V, E)$ and a sequence $v^1, v^2, \ldots, v^r$ of vertices in G. A **path** is a sequence of edges $e_1, e_2, \ldots, e_{r-1}$ in E such that $e_i = (v^i, v^{i+1})$. The vertex v^1 is the initial vertex on the path and v^r is the terminal vertex. An example of a path is the sequence $(1, 2), (2, 3), (3, 4)$ in Figure 1.1. A *cycle* is a path whose initial and terminal vertices are the same. The edges $(2, 3), (3, 4), (2, 4)$ form a cycle in Figure 1.1.

A graph G is called **connected** if there is a path in G between every pair of vertices. Figure 1.2(a) shows a connected graph while Figure 1.2(b) shows a disconnected one. If G is a connected graph then $T \subset E$ is called **acyclic** or a **forest** if (V, T) contains no cycles. The set $\{(1, 2), (3, 4)\}$ in Figure 1.1 is a forest. If in addition (V, T) is connected then T is called a **spanning tree**. The set $\{(1, 2), (2, 3), (3, 4)\}$ is a spanning tree in Figure 1.1. It is easy to see that every connected graph contains a spanning tree.

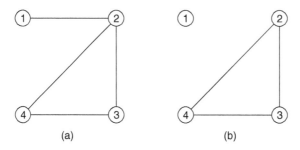

(a)   (b)

Figure 1.2

Lemma 1.27 *Let $G = (V, E)$ be a connected graph and $T \subseteq E$ a spanning tree of G. Then, (V, T) contains at least one vertex of degree 1.*

Proof Suppose not. We will select a path in G and show it to be a cycle. Initially all vertices are classified as unmarked. Select any vertex $v \in V$ and mark it zero. Find an adjacent vertex that is unmarked and mark it 1. Repeat, each time marking a vertex with a number one higher than the last marked vertex. If there are no unmarked vertices to choose from, stop. Since there are a finite number of vertices, this marking procedure must end. Since every vertex has degree at least 2, the last vertex to be marked, with mark k, say is adjacent to at least one vertex with a mark $k - 2$ or smaller, say, r. The path determined by the vertices with marks $r, r + 1, \ldots, k - 1, k$ forms a cycle, a contradiction. ∎

Theorem 1.28 *Let $G = (V, E)$ be a connected graph and $T \subseteq E$ a spanning tree of G. Then $|T| = |V| - 1$.*

Proof The proof is by induction on $|T|$. If $|T| = 1$, since (V, T) is connected it follows that V must have just two elements, the endpoints of the lone edge in T. Now suppose the lemma is true whenever $|T| \leq m$. Consider an instance where

$|T| = m + 1$. Let $u \in V$ be the vertex in (V, T) with degree one. Such a vertex exists by the previous lemma. Let $v \in V$ be such that $(u, v) \in T$. Notice that $T \setminus (u, v)$ is a spanning tree for $(V \setminus u, E)$. By induction, $|T \setminus (u, v)| = |V| - 2$, therefore $|T| = |V| - 1$. ■

Since a tree T is connected it contains at least one path between every pair of vertices. In fact it contains a unique path between every pair of vertices. Suppose not. Then there will be at least two edge disjoint paths between some pair of vertices. The union of these two paths would form a cycle, contradicting the fact that T is a tree.

1.5.1   Directed graphs

If the edges of a graph are oriented, i.e., an edge (i, j) can be traversed from i to j but not the other way around, the graph is called **directed**. If a graph is directed, the edges are sometimes called **arcs**. Formally, a directed graph consists of a set V of vertices and set E of **ordered** pairs of vertices. As an example, suppose $V = \{1, 2, 3, 4\}$ and $E = \{(1, 2), (2, 3), (3, 4), (2, 4), (4, 2), (4, 1)\}$. A pictorial representation of this graph is shown in Figure 1.3. A path in a directed graph has the same definition as in the undirected case except now the orientation of each edge must be respected. To emphasize this it is common to call a path directed. In our example above, $1 \rightarrow 4 \rightarrow 3$ would not be a directed path, but $1 \rightarrow 2 \rightarrow 4$ would be. A cycle in a directed graph is defined in the same way as in the undirected case, but again the orientation of the edges must be respected.

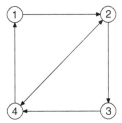

Figure 1.3

A directed graph is called **strongly connected** if there is a directed path between every ordered pair of vertices. It is easy to see that this is equivalent to requiring that there be a directed cycle through every pair of vertices.

Problems

1.1 Show that the function $f(x) = |x|$ is continuous for $x \in \mathbb{R}^1$. Now show that $g(x) = (1 + |x|)^{-1}$ is continuous for $x \in \mathbb{R}^1$.

1.2 Show that any polynomial function of $x \in \mathbb{R}^1$ is continuous.

1.3 Show that the rank of an $m \times n$ matrix is at most $\min\{m, n\}$.

1.4 Compute the rank of the following matrix:

$$\begin{bmatrix} 2 & 0 & -1 & 3 \\ 1 & 1 & 2 & 2 \\ 2 & 0 & -1 & 1 \end{bmatrix}.$$

1.5 Let A be an $m \times n$ matrix of rank r and $b \in \mathbb{R}^m$. Let r' be the rank of the augmented matrix $[A|b]$ and let $F = \{x \in \mathbb{R}^n : Ax = b\}$. Prove that exactly one of the following must be true:

1.   if $r' = r + 1$, $F = \varnothing$;
2.   if $r' = r = n$, F is a single vector;
3.   if $r = r' < n$, then F contains infinitely many vectors of the form $y + z$ where $Ay = b$ and z is in the kernel of A.

1.6 A house has many rooms and each room has either 0, 1 or 2 doors. An **outside** door is one that leads out of the house. A room with a single door is called a **dead end**. Show that the number of dead ends must have the same parity as the number of outside doors.

Note

1 Named after Augustin Louis Cauchy (1789–1857). Actually it had been discovered four years earlier by Bernard Bolzano (1781–1848).

References

Bollobás, B.: 1979, *Graph theory: an introductory course*, Graduate texts in mathematics; 63, Springer Verlag, New York.
Clapham, C. R. J.: 1973, *Introduction to mathematical analysis*, Routledge & K. Paul, London, Boston.
Lang, S.: 1987, *Linear algebra*, Undergraduate texts in mathematics, 3rd edn, Springer-Verlag, New York.
Simon, C. P. and Blume, L.: 1994, *Mathematics for economists*, 1st edn, Norton, New York.

2 Feasibility

Let A be an $m \times n$ matrix of real numbers. We will be interested in problems of the following kind:

> Given $b \in \mathbb{R}^m$ find an $x \in \mathbb{R}^n$ such that $Ax = b$ or prove that no such x exists.

Convincing another that $Ax = b$ has a solution (when it does) is easy. One merely exhibits the solution and they can verify that the solution does indeed satisfy the equations. What if the system $Ax = b$ does not admit a solution? Is there an easy way to convince another of this? Stating that one has checked all possible solutions is not persuasive; there are infinitely many.

By framing the problem in the right way we can bring to bear the machinery of linear algebra. Specifically, given $b \in \mathbb{R}^m$, the problem of finding an $x \in \mathbb{R}^n$ such that $Ax = b$ can be stated as: is $b \in \text{span}(A)$?

2.1 Fundamental theorem of linear algebra

Suppose we wish to know if the following system has a solution:

$$\begin{bmatrix} -4 & 2 & -5 \\ 2 & -1 & 2.5 \end{bmatrix} \begin{bmatrix} x_1 \\ x_2 \\ x_3 \end{bmatrix} = \begin{bmatrix} 1 \\ 1 \end{bmatrix}.$$

For the moment suppose that it does, call it, x^*. Adding a linear multiple of one equation to the other yields another equation that x^* must also satisfy. Multiply the second equation by 2 and add it to the first. This produces

$$0 \times x_1 + 0 \times x_2 + 0 \times x_3 = 3$$

which clearly has no solution. Therefore the original system cannot have a solution. Our manipulation of the equations has produced an inconsistency which certifies that the given system is insoluble. It suggests that we might be able to decide the insolvability of a system by deriving, through appropriate linear combinations of the given equations, an inconsistency. That this is possible was first proved by Gauss.[1]

Theorem 2.1   *Let A be an m × n matrix, b ∈ ℝ^m and F = {x ∈ ℝ^n: Ax = b}. Then either F ≠ ∅ or there exists y ∈ ℝ^m such that yA = 0 and yb ≠ 0 but not both.*

Remark   Suppose $F = ∅$. Then, b is not in the span of the columns of A. If we think of the span of the columns of A as a plane, then b is a vector pointing out of the plane (see Figure 2.1). Thus, any vector, y orthogonal to this plane (and so to every column of A) must have a non-zero dot product with b. Now for an algebraic interpretation. Take any linear combination of the equations in the system $Ax = b$. This linear combination can be obtained by pre-multiplying each side of the equation by a suitable vector y, i.e., $yAx = yb$. Suppose there is a solution x^* to the system, i.e., $Ax^* = b$. Any linear combination of these equations results in an equation that x^* satisfies as well. Consider the linear combination obtained by multiplying the ith equation through by y_i and summing over the row index i. In particular, x^* must also be a solution to the resulting equation: $yAx = yb$. Suppose we found a vector y such that $yAx ≠ yb$ then clearly the original system $Ax = b$ could not have a solution.

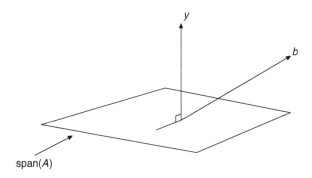

Figure 2.1

Proof   First we prove the 'not both' part. Suppose $F ≠ ∅$. Choose any $x ∈ F$. Then

$$yb = yAx = (yA)x = 0$$

which contradicts the fact that $yb ≠ 0$.

If $F ≠ ∅$ we are done. Suppose that $F = ∅$. Hence b cannot be in the span of the columns of A. Thus the rank of $C = [A|b]$, r', is one larger than the rank, r, of A. That is, $r' = r + 1$.

Since C is a $m \times (n + 1)$ matrix,

$$\text{rank}(C^T) + \dim[\ker(C^T)] = m = \text{rank}(A^T) + \dim[\ker(A^T)].$$

Using the fact that the rank of a matrix and its transpose coincide we have

$$r' + \dim[\ker(C^T)] = r + \dim[\ker(A^T)],$$

i.e. $\dim[\ker(C^T)] = \dim[\ker(A^T)] - 1$. Since the dimension of $\ker(C^T)$ is one smaller than the dimension of $\ker(A^T)$ we can find a $y \in \ker(A^T)$ that is not in $\ker(C^T)$. Hence $yA = 0$ but $yb \neq 0$. ∎

2.2 Linear inequalities

Now consider the following problem:

> Given a $b \in \mathbb{R}^m$ find an $x \in \mathbb{R}^n$ such that $Ax \leq b$ or show that no such x exists.

The problem differs from the earlier one in that '=' has been replaced by '$\leq$'. We deal first with a special case of this problem and then show how to reduce the problem above to this special case.

2.3 Non-negative solutions

We focus on finding a **non-negative** $x \in \mathbb{R}^n$ such that $Ax = b$ or show that no such x exists. Observe that if $b = 0$, the problem is trivial, so we assume that $b \neq 0$. The problem can be framed as follows: can b be expressed as a non-negative linear combination of the columns of A?

Definition 2.2 *A set C of vectors is called a* **cone** *if $\lambda x \in C$ whenever $x \in C$ and $\lambda > 0$.*

For example, the set $\{(x_1, 0): x_1 \geq 0\} \cup \{(0, x_2): x_2 \geq 0\}$ is a cone. A special class of cones that will play an important role is defined next. The reader should verify that the set so defined is a cone.

Definition 2.3 *The set of all non-negative linear combinations of the columns of A is called the* **finite cone** *generated by the columns of A. It is denoted* cone(A).

Example 5 *Suppose A is the following matrix:*

$$\begin{bmatrix} 2 & 0 \\ 1 & 1 \end{bmatrix}$$

The cone generated by the columns of A is the shaded region in Figure 2.2.

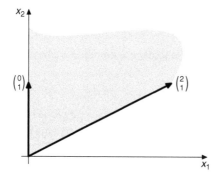

Figure 2.2

The reader should compare the definition of span(A) with cone(A). In particular

$$\text{span}(A) = \{y \in \mathbb{R}^m : \text{s.t. } y = Ax \text{ for some } x \in \mathbb{R}^n\}$$

and

$$\text{cone}(A) = \{y \in \mathbb{R}^m : \text{s.t. } y = Ax \text{ for some } x \in \mathbb{R}_+^n\}.$$

To see the difference consider the matrix

$$\begin{bmatrix} 1 & 0 \\ 0 & 1 \end{bmatrix}.$$

The span of the columns of this matrix will be all of $\mathbb{R}^2$, while the cone generated by its columns is the non-negative orthant.

Theorem 2.4 (Farkas Lemma)[2]   *Let A be an $m \times n$ matrix, $b \in \mathbb{R}^m$ and $F = \{x \in \mathbb{R}^n : Ax = b, x \geq 0\}$. Then either $F \neq \varnothing$ or there exists $y \in \mathbb{R}^m$ such that $yA \geq 0$ and $y \cdot b < 0$ but not both.*

Remark   Like the fundamental theorem of linear algebra, this result is capable of a geometric interpretation which is deferred to a later chapter. The algebraic interpretation is this. Take any linear combination of the equations in $Ax = b$ to get $yAx = yb$. A non-negative solution to the first system is a solution to the second. If we can choose y so that $yA \geq 0$ and $y \cdot b < 0$, we find that the left hand side of the single equation $yAx = yb$ is at least zero while the right hand side is negative, a contradiction. Thus the first system cannot have a non-negative solution.

Proof   First we prove that both statements cannot hold simultaneously. Suppose not. Let $x^* \geq 0$ be a solution to $Ax = b$ and y^* a solution to $yA \geq 0$ such that

$y^*b < 0$. Notice that x^* must be a solution to $y^*Ax = y^*b$. Thus $y^*Ax^* = y^*b$. Then $0 \le y^*Ax^* = y^*b < 0$, a contradiction.

If $b \notin \text{span}(A)$, by Theorem 2.1 there is a $y \in \mathbb{R}^m$ such that $yA = 0$ and $yb \ne 0$. If it so happens that the given y has the property that $yb > 0$ we are done. If $yb < 0$, then negate y and again we are done. So, we may suppose that $b \in \text{span}(A)$ but $b \notin \text{cone}(A)$, i.e. $F = \varnothing$.

Let r be the rank of A. Note that $n \ge r$. Since A contains r LI column vectors and $b \in \text{span}(A)$, we can express b as a linear combination of an r-subset D of LI columns of A. Let $D = \{a^{i_1}, \ldots, a^{i_r}\}$ and $b = \sum_{t=1}^{r} \lambda_{i_t} a^{i_t}$. Note that D is LI. Since $b \notin \text{cone}(A)$, at least one of $\{\lambda_{i_t}\}_{t \ge 1}$ is negative.

Now apply the following four step procedure repeatedly. Subsequently, we show that the procedure must terminate.

1. Choose the smallest index h amongst $\{i_1, \ldots, i_r\}$ with $\lambda_h < 0$.
2. Choose y so that $y \cdot a = 0$ for all $a \in D \setminus a^h$ and $y \cdot a^h \ne 0$. This can be done by Theorem 2.1 because $a^h \notin \text{span}(D \setminus a^h)$. Normalize y so that $y \cdot a^h = 1$. Observe that $y \cdot b = \lambda_h < 0$.
3. If $y \cdot a^j \ge 0$ for all columns a^j of A stop, and the proof is complete.
4. Otherwise, choose the smallest index w such that $y \cdot a^w < 0$. Note that $w \notin D \setminus a^h$. Replace D by $\{D \setminus a^h\} \cup a^w$, i.e., exchange a^h for a^w.

To complete the proof, we must show that the procedure terminates (see step 3).[3] Let D^k denote the set D at the start of the kth iteration of the four step process described above. If the procedure does not terminate there is a pair $k < l$ such that $D^k = D^l$, i.e., the procedure cycles.

Let s be the largest index for which a^s has been removed from D at the end of one of the iterations $k, k+1, \ldots, l-1$, say p. Since $D^l = D^k$ there is a q such that a^s is inserted into D^q at the end of iteration q, where $k \le q < l$. No assumption is made about whether $p < q$ or $p > q$. Notice that

$$D^p \cap \{a^{s+1}, \ldots, a^n\} = D^q \cap \{a^{s+1}, \ldots, a^n\}.$$

Let $D^p = \{a^{i_1}, \ldots, a^{i_r}\}$, $b = \lambda_{i_1} a^{i_1} + \cdots + \lambda_{i_r} a^{i_r}$ and let y' be the vector found in step two of iteration q. Then:

$$0 > y' \cdot b = y'(\lambda_{i_1} a^{i_1} + \cdots + \lambda_{i_r} a^{i_r}) = y'\lambda_{i_1} a^{i_1} + \cdots + y'\lambda_{i_r} a^{i_r} > 0,$$

a contradiction. To see why the last inequality must be true:

*   When $i_j < s$, we have from step 1 of iteration p that $\lambda_{i_j} \ge 0$. From step 4 of iteration q we have $y' \cdot a^{i_j} \ge 0$.
*   When $i_j = s$, we have from step 1 of iteration p that $\lambda_{i_j} < 0$. From step 4 of iteration q we have $y' \cdot a^{i_j} < 0$.

- When $i_j > s$, we have from $D^p \cap \{a^{s+1}, \ldots, a^r\} = D^q \cap \{a^{r+1}, \ldots, a^r\}$ and step 2 of iteration q that $y' \cdot a^{i_j} = 0$.

This completes the proof.   ∎

This particular proof is a disguised form of the simplex algorithm developed by George Dantzig (1914–). Dantzig we will meet again. The particular way of choosing which elements to enter or leave the set D is due to Robert Bland and called Bland's anti-cycling rule .

In fact, we have proven more. Suppose $b \notin \text{cone}(A)$ and A has rank r. Then there are r LI columns of A, $\{a^1, a^2, \ldots, a^{r-1}\}$ and $y \in \mathbb{R}^m$ such that $y \cdot a^j = 0$ for $1 \leq j \leq r - 1$, $y \cdot a^j \geq 0$ for all $j \geq r$ but $y \cdot b < 0$. This fact will be useful later. The system $yA \geq 0$ and $yb < 0$ is sometimes referred to as the **Farkas alternative**. It is useful to recall that the Farkas lemma can also be stated this way:

Either $yA \geq 0$, $y \cdot b < 0$ has a solution or $Ax = b, x \geq 0$ has a solution **but not both.**

Example 6   *We use the Farkas lemma to decide if the following system has a non-negative solution:*

$$\begin{bmatrix} 4 & 1 & -5 \\ 1 & 0 & 2 \end{bmatrix} \begin{bmatrix} x_1 \\ x_2 \\ x_3 \end{bmatrix} = \begin{bmatrix} 1 \\ 1 \end{bmatrix}.$$

The Farkas alternative is

$$\begin{bmatrix} y_1 & y_2 \end{bmatrix} \begin{bmatrix} 4 & 1 & -5 \\ 1 & 0 & 2 \end{bmatrix} \geq \begin{bmatrix} 0 \\ 0 \end{bmatrix},$$

$$y_1 + y_2 < 0.$$

As a system of inequalities the alternative is:

$$4y_1 + y_2 \geq 0,$$
$$y_1 + 0y_2 \geq 0,$$
$$-5y_1 + 2y_2 \geq 0,$$
$$y_1 + y_2 < 0.$$

There can be no solution to this system. The second inequality requires that $y_1 \geq 0$. Combining this with the the last inequality we conclude that $y_2 < 0$. But $y_1 \geq 0$ and $y_2 < 0$ contradict $-5y_1 + 2y_2 \geq 0$. So, the original system has a non-negative solution.

Example 7 *We use the Farkas lemma to decide the solvability of the system:*

$$\begin{bmatrix} 1 & 1 & 0 \\ 0 & 1 & 1 \\ 1 & 0 & 1 \\ 1 & 1 & 1 \end{bmatrix} \begin{bmatrix} x_1 \\ x_2 \\ x_3 \end{bmatrix} = \begin{bmatrix} 2 \\ 2 \\ 2 \\ 1 \end{bmatrix}.$$

We are interested in non-negative solutions of this system. The Farkas alternative is

$$\begin{bmatrix} y_1 & y_2 & y_3 & y_4 \end{bmatrix} \begin{bmatrix} 1 & 1 & 0 \\ 0 & 1 & 1 \\ 1 & 0 & 1 \\ 1 & 1 & 1 \end{bmatrix} \geq \begin{bmatrix} 0 \\ 0 \\ 0 \\ 0 \end{bmatrix},$$

$$2y_1 + 2y_2 + 2y_3 + y_4 < 0.$$

One solution is $y_1 = y_2 = y_3 = -1/2$ *and* $y_4 = 1$, *implying that the given system has no solution.*

In fact, the solution to the alternative provides an 'explanation' for why the given system has no solution. Multiply each of the first three equations by $1/2$ *and add them together to yield*

$$x_1 + x_2 + x_3 = 3.$$

However, this is inconsistent with the fourth equation which reads $x_1 + x_2 + x_3 = 1$.

2.4 The general case

The problem of deciding whether the system $\{x \in \mathbb{R}^n \colon Ax \leq b\}$ has a solution can be reduced to the problem of deciding if $Bz = b$, $z \geq 0$ has a solution for a suitable matrix B.

First observe that any inequality of the form $\sum_j a_{ij}x_j \geq b_i$ can be turned into an equation by the subtraction of a **surplus variable**, s. That is, define a new variable $s_i \geq 0$ such that

$$\sum_j a_{ij}x_j - s_i = b_i.$$

Similarly, an inequality of the form $\sum_j a_{ij}x_j \leq b_i$ can be converted into an equation by the addition of a **slack variable**, $s_i \geq 0$ as follows:

$$\sum_j a_{ij}x_j + s_i = b_i.$$

A variable, x_j that is unrestricted in sign can be replaced by two non-negative variables z_j and z'_j by setting $x_j = z_j - z'_j$. In this way any inequality system can be converted into an equality system with non-negative variables. We will refer to this as converting into standard form.

As an example we derive the Farkas alternative for the system $\{x: Ax \leq b, x \geq 0\}$. Deciding solvability of $Ax \leq b$ for $x \geq 0$ is equivalent to solvability of $Ax + Is = b$ where $x, s \geq 0$. Set $B = [A|I]$ and $z = \binom{x}{s}$ and we can write the system as $Bz = b, z \geq 0$. Now apply the Farkas lemma to this system:

$$yB \geq 0, \quad yb < 0.$$

Now $0 \leq yB = y[A|I]$ implies $yA \geq 0$ and $y \geq 0$. So, the Farkas alternative is $\{y: yA \geq 0, y \geq 0, yb < 0\}$.

The principle here is that by a judicious use of auxiliary variables one can convert almost anything into standard form.

2.5 Application: arbitrage

The word *arbitrage* comes from the French *arbitrer* and means to trade in stocks in different markets to take advantage of different prices. H. R. Varian (1987) offers the following story to illustrate arbitrage:

> An economics professor and Yankee farmer were waiting for a bus in New Hampshire. To pass the time, the farmer suggested that they play a game. "What kind of game would you like to play?" responded the professor. "Well," said the farmer, "how about this: I'll ask a question, and if you can't answer my question, you give me a dollar. Then you ask me a question and if I can't answer your question, I'll give you a dollar."
>
> "That sounds attractive," said the professor, "but I do have to warn you of something: I'm not just an ordinary person. I'm a professor of economics."
>
> "Oh," replied the farmer, "In that case we should change the rules. Tell you what: if you can't answer my question you still give me a dollar, but if I can't answer yours, I only have to give you fifty cents."
>
> "Yes," said the professor, "that sounds like a fair arrangement."
>
> "Okay," said the farmer, "Here's my question: what goes up the hill on seven legs and down on three legs?"
>
> The professor pondered this riddle for a little while and finally replied. "Gosh, I don't know ... what does go up the hill on seven legs and down on three legs?"
>
> "Well," said the farmer, "I don't know either. But if you give me your dollar, I'll give you my fifty cents!"

The absence of arbitrage opportunities is the driving principle of financial theory.[4]

Arguments relying on the absence of arbitrage made their appearance in finance in the 1970s, but they are much older. In the early 1920s Frank Ramsey[5] outlined a definition of probability based on the absence of arbitrage.[6] In 1937, Bruno de Finetti[7] (1906–1985), independently, used the absence of arbitrage as a basis for defining subjective probability. This paper had the ironic fate to stimulate the very ideas (subjective expected utility[8]) that were to outshine it.

de Finetti proposed a definition of probability in terms of prices placed on lottery tickets. Let $p(E)$ be the unit price at which one would be indifferent between buying and selling a lottery ticket that paid \$1 if event E occurred and 0 otherwise. Let $p(E|F)$ be the unit price at which one would be indifferent between buying and selling a ticket paying \$1 if $E \cap F$ occurs, 0 if F occurs without E and a refund of the purchase price if F fails to occur. de Finetti showed that such a system of prices eliminates arbitrage if and only if the prices satisfied the requirements of a probability measure. That is,

- $p(E) \geq 0$,
- $p(E) + p(E^c) = 1$,
- $p(E \cup F) = p(E) + p(F)$ if $E \cap F = \varnothing$,
- $p(E \cap F) = p(E|F)p(F)$.

On this basis de Finetti argued that probabilities should be interpreted as these prices. Indeed, he argued that probability had no meaning beyond this. In the preface of his 1974 book entitled 'Theory of Probability' he writes:

> "PROBABILITY DOES NOT EXIST. The abandonment of superstitious beliefs about the existence of Phlogiston, the Cosmic Ether, Absolute Space and Time . . . or Fairies and Witches, was an essential step along the road to scientific thinking. Probability, too, if regarded as something endowed with some kind of objective existence, is no less a misleading conception, an illusory attempt to exteriorize or materialize our true probabilistic beliefs."

Here we recast de Finetti's theorem in a form that is useful for Finance applications. Suppose m assets each of whose payoffs depends on a future state of nature. Let S be the set of possible future states of nature with $n = |S|$. Let a_{ij} be the payoff from one share of asset i in state j. A portfolio of assets is represented by a vector $y \in \mathbb{R}^m$ where the ith component, y_i represents the amount of asset i held. If $y_i > 0$, one holds a **long** position in asset i while $y_i < 0$ implies a **short** position in asset i.[9] Let $w \in \mathbb{R}^n$ be a vector whose jth component denotes wealth in state $j \in S$. We assume that wealth (w) in a future state is related to the current portfolio (y) by

$$w = yA.$$

This assumes that assets are infinitely divisible, returns are linear in the quantities held and the return of the asset is not affected by whether one holds a long or short position. Thus, if one can borrow from the bank at 5% one can lend to the bank at 5%.

The no arbitrage condition asserts that a portfolio that pays off non-negative amounts in every state must have a non-negative cost. If $p > 0$ is a vector of asset prices, we can state the no arbitrage condition algebraically as follows:

$$yA \geq 0 \quad \Rightarrow \quad y \cdot p \geq 0.$$

Equivalently, the system $yA \geq 0, y \cdot p < 0$ has no solution. From the Farkas lemma we deduce the existence of a non-negative vector $\hat{\pi} \in \mathbb{R}^m$ such that

$$p = A\hat{\pi}.$$

Since $p > 0$, it follows that $\hat{\pi} > 0$. Scale $\hat{\pi}$ by dividing through by $\sum_j \hat{\pi}_j$. Let $p^* = p / \sum_j \hat{\pi}_j$ and $\pi = \hat{\pi} / \sum_j \hat{\pi}_j$. Notice that π is a probability vector. As long as relative prices are all that matter, scaling the prices is of no relevance. After the scaling, $p^* = A\pi$. In words, there is a probability distribution under which every securities expected value is equal to its buying/selling price. Such a distribution is called a **risk neutral probability distribution**. A risk-neutral investor using these probabilities would conclude that the securities are fairly priced.

In this set up, the market is said to be **complete** if $\text{span}(A) = \mathbb{R}^m$. If a market is complete and has more assets (m) than states of nature (n), some of the assets will be redundant. The payoffs of the redundant assets can be duplicated by a suitable portfolio of other assets. In this case it is usual to restrict oneself to a subset of the securities that form a basis for the row space of A.[10] When $m < n$, the market is said to be **incomplete** because there can be a wealth vector w not attainable by any portfolio y, i.e., given w there is no y such that $w = yA$.

2.5.1 Black–Scholes formula

The most remarkable application of the arbitrage idea is to the pricing of derivative securities called options. The idea is due to Fischer Black, Myron Scholes and Robert Merton. A **call option** on a stock is a contract giving one the 'option' to buy the stock at a specified price (**strike price**) at a specified time in the future.[11] The advantage of a call option is that it allows one to postpone the purchase of a stock until after one sees the price. In particular one can wait for the market price of the stock to rise above the strike price. Then, exercise the option. That is, buy at the strike price and resell at the higher market price. How much is a call option worth?

For simplicity, assume a single time period and a market consisting of a stock, a bond and a call option on the stock. Let K be the strike price of the call option and suppose that it can be exercised only at the end of the time period.[12] Suppose S is the value of the stock at the end of the time period. If $S > K$, the option holder will exercise the call option yielding a profit of $S - K$ dollars. If $S \leq K$, the option holder will let the call option expire and its value to the holder is zero. In short the call option has a payoff of max$\{0, S - K\}$.

Suppose that there are only two possible future states of the world (good, bad). The good state is where the stock goes up by a factor $u > 1$. The bad state is where the stock declines by a factor $d < 1$. Investing in the bond is risk free. This means that in all cases in the future the value of the bond goes up by a factor $r > 1$. Let S^0 be the initial stock price and B the price of the bond.

We now have an economy with three assets (stock, bond, call option) and two possible states of the world. Let a_{ij} be the value of asset i in state j. For our example, the matrix $A = \{a_{ij}\}$ will be

$$
\begin{bmatrix}
u S^0 & d S^0 \\
r B & r B \\
\max\{0, u S^0 - K\} & \max\{0, d S^0 - K\}
\end{bmatrix}.
$$

The first column corresponds to the good state, the second to the bad state. The first row corresponds to the stock, the second to the bond and the third to the call option. If p is the vector of prices then $p_1 = S^0$, $p_2 = B$ and p_3 is the price of the call option we wish to determine.

The absence of arbitrage implies the existence of π_1, π_2 non-negative such that

$$
\begin{bmatrix}
u S^0 & d S^0 \\
r B & r B \\
\max\{0, u S^0 - K\} & \max\{0, d S^0 - K\}
\end{bmatrix}
\begin{bmatrix}
\pi_1 \\
\pi_2
\end{bmatrix}
=
\begin{bmatrix}
S^0 \\
B \\
p_3
\end{bmatrix}.
$$

Consider the first two equations:

$$
u S^0 \pi_1 + d S^0 \pi_2 = S^0,
$$

$$
r B \pi_2 + r B \pi_2 = B.
$$

They have the unique solution

$$
\pi_1 = \frac{r - d}{r(u - d)}, \quad \pi_2 = \frac{u - r}{r(u - d)}.
$$

This solution is non-negative iff $u > r > d$. Does it make sense to impose these conditions at the outset? Observe that if $r \leq d$, the bond would be worthless; i.e., one would always better off investing in the stock. If $u \leq r$ no one would be interested in buying the stock.

Using this solution we deduce that

$$
p_3 = \pi_1 \max\{0, u S^0 - K\} + \pi_2 \max\{0, d S^0 - K\}.
$$

Notice that it does not depend on the probabilities of either the good or bad state being realized. This option pricing formula is the discrete (one period) analogue of the famous Black–Scholes formula.

2.6   Application: co-operative games

A co-operative game (with transferable utility) is defined by a set N of players and a value function $v\colon 2^N \to \mathbb{R}$ which represents the monetary value or worth of a subset S of players forming a coalition. The story here is that if the set S of players were to pool their resources and use them appropriately, they would generate $v(S)$ dollars to be consumed by themselves. The value of $v(S)$ tells us nothing about how it is to be divided amongst the players.

It is usual to assume that $v(N) \geq \max_{S \subset N} v(S)$. That is, the largest possible value is generated if all players work together. We will be interested in how to apportion $v(N)$ between the players so as to give each player an incentive to combine into a whole.

A vector $x \in \mathbb{R}^n$ is called an **imputation** if $\sum_{j \in N} x_j = v(N)$ and $x_j \geq v(j)$ for all $j \in N$. One can think of an imputation as a division of $v(N)$ that gives to every player at least as much as they could get by themselves. One can require that the division satisfy a stronger requirement. Specifically, every subset S of agents should receive in total at least as much as $v(S)$. This leads us to the notion of core.

Definition 2.5   *The* **core** *of the game* (v, N) *is the set*

$$C(v, N) = \left\{ x \in \mathbb{R}^n : \sum_{j \in N} x_j = v(N), \sum_{j \in S} x_j \geq v(S), \forall S \subset N \right\}.$$

Example 8   *Suppose* $N = \{1, 2, 3\}$, $v(\{1\}) = v(\{2\}) = v(\{3\}) = 0$, $v(\{1, 2\}) = v(\{2, 3\}) = v(\{1, 3\}) = 2$ *and* $v(N) = 2.9$. *The core is the set of solutions to the following:*

$$
\begin{aligned}
x_1 \;+\; x_2 \;+\; x_3 &= 2.9, \\
x_1 \;+\; x_2 \phantom{{}+{}x_3} &\geq 2, \\
x_1 \phantom{{}+{}x_2} \;+\; x_3 &\geq 2, \\
\phantom{x_1 \;+{}} x_2 \;+\; x_3 &\geq 2, \\
x_1, x_2, x_3 &\geq 0.
\end{aligned}
$$

If we add up the second, third and fourth inequality we deduce that $x_1 + x_2 + x_3 \geq 3$ which contradicts the first equation. Therefore the core is empty.

Let $B(N)$ be the set of feasible solutions to the following system:

$$\sum_{S : i \in S} y_S = 1, \quad \forall i \in N,$$

$$y_S \geq 0, \quad \forall S \subset N.$$

The reader should verify that $B(N) \neq \varnothing$.

Theorem 2.6 (Bondareva–Shapley) $\quad C(v, N) \neq \varnothing$ *iff*

$$v(N) \geq \sum_{S \subset N} v(S) y_S, \quad \forall y \in B(N).$$

Proof The Farkas alternative for the system that defines $C(v, N)$ is

$$v(N) - \sum_{S \subset N} v(S) y_s < 0,$$

$$\sum_{S: i \in S} y_S = 1, \quad \forall i \in N,$$

$$y_S \geq 0, \quad \forall S \subset N.$$

By the Farkas lemma, $C(v, N) \neq \varnothing$ iff the alternative is infeasible. Since $B(N) \neq \varnothing$, infeasibility of the alternative implies that for all $y \in B(N)$ we have $v(N) - \sum_{S \subset N} v(S) y_s \geq 0$ from which the result follows. ∎

2.7 Application: auctions

Auctions are a venerable and popular selling institution. The word auction comes from the Latin *auctus* meaning to increase. An even obscurer term for auction is the Latin word *subhastare*. It is the conjuction of *sub* meaning 'under' and *hasta* meaning 'spear'. After a military victory a Roman soldier would plant his spear in the ground to mark the location of his spoils. Later he would put these goods up for sale by auction.[13]

Perhaps the most engaging tale about auctions that no writer can decline telling is the sale of the Roman empire to the highest bidder. It is described in Edward Gibbon's account of the decline and fall of the same.[14]

In 193 A.D. the Praetorian guard[15] killed the emperor Pertinax.[16] Sulpicanus, father in law to Pertinax offered the Praetorians 5,000 drachmas per guard to be emperor. Realizing they were onto a good thing, the guard announced that the Empire was available for sale to the highest bidder. Didius Julianus outbid all comers and became the emperor for the price of 6,250 drachmas per Guard.[17] He was beheaded two months later when Septimus Severus conquered Rome.

We illustrate how the Farkas lemma can be used in the design of an auction.[18] An auction can take various forms but for our purposes it consists of two steps. In the first, bidders announce how much they are willing to pay (bids) for the object. In the second, the seller chooses, in accordance with a previously announced function of the bids, who gets the object and how much each bidder must pay. This choice could be random.[19]

The simplest set up involves two risk neutral bidders and one seller. The seller does not know how much each bidder is willing to pay for the object. Each bidder is ignorant of the others valuations. It is the uncertainty about valuations that makes auctions interesting objects of study. If the seller knew the valuations, she would approach the bidder with the highest valuation and make him a take it or leave it offer slightly under their valuation and be done with.

The uncertainty in bidder valuations is typically modeled by assuming that their monetary valuations are drawn from a commonly known distribution over a finite set W.[20]

In some contexts it is natural to suppose that the valuations are drawn independently. This captures the idea of values being purely subjective. The value that one bidder enjoys from the consumption of the good does not influence the value that the other bidders will enjoy. Here we suppose that the valuations are correlated. One context where such an assumption makes sense is in bidding for oil leases. The value of the lease depends on the amount of oil under the ground. Each bidders estimate of that value depends on seismic and other surveys of the land in question. It is reasonable to suppose that one bidders survey results would be correlated with anothers because they are surveying the same plot of land.

Denote by v^i the value that bidder i places on the object. For any two $a, b \in W$ let $p_{ab} = \Pr[v^2 = b | v^1 = a] = \Pr[v^1 = b | v^2 = a]$. The important assumption we make is that no row of the matrix $\{p_{ab}\}$ is a non-negative linear combination of the other rows. We refer to this as the **cone assumption**. Were the values drawn independently, the rows of this matrix would be identical.

Each bidder is asked to report their value. Let T^1_{ab} be the payment that bidder 1 makes if he reports a and bidder 2 reports b. Similarly define T^2_{ab}. Let Q^1_{ab} be the probability that the object is assigned to agent 1 when he reports a and bidder 2 reports b. Notice that $Q^2_{ab} = 1 - Q^1_{ab}$.

Two constraints are typically imposed on the auction design. The first is called **incentive compatibility**. The expected payoff to each bidder from reporting truthfully (assuming the other does so as well) should exceed the expected payoff from bidding insincerely. Supposing bidder 1's valuation for the object is a, this implies that

$$\sum_{b \in W} p_{ab}[Q^1_{ab}a - T^1_{ab}] \geq \sum_{b \in W} p_{ab}[Q^1_{kb}a - T^1_{kb}] \quad \forall k \in W \setminus a.$$

The left-hand side of this inequality is the expected payoff (assuming the other bidder reports truthfully) to a bidder with value a who reports a. The right hand side is the expected payoff (assuming the other bidder reports truthfully) to a bidder with value a who reports k as their value. This constraint must hold for each $a \in W$ and a similar one must hold for bidder 2.

The incentive compatibility constraint does not force any bidder to bid sincerely. Only if all other bidders bid sincerely, is it the case that one should bid sincerely. Furthermore, the inequality in the incentive compatibility constraint means that it is possible for a bidder to be indifferent between bidding sincerely or lying. At best the incentive compatibility constraint ensures that bidding sincerely is in a sense mutually rational. One could demand that the auction design offer greater incentives to bid sincerely than the ones considered here, but that is a subject for another book.

The second constraint, called **individual rationality**, requires that no bidder should be made worse off by participating in the auction. It is not obvious how to express this constraint as an inequality, since the act of participation does not tell us how a bidder will bid. This is where the incentive compatibility constraint is useful. With it we can argue that if a bidder participates, she will do so by bidding sincerely. Hence, if bidder 1's valuation is $a \in W$ and he reports this, which follows from incentive compatibility, we can express individual rationality as:

$$\sum_{b \in W} p_{ab}[Q^1_{ab}a - T^1_{ab}] \geq 0.$$

This constraint must hold for each $a \in W$ and for bidder 2 as well.

The goal of the auctioneer is to design the auction so as to maximize her expected revenue subject to incentive compatibility and individual rationality. Notice that her expected revenue is maximized when the expected profit to all bidders is 0. Given incentive compatibility, bidder 1's expected profit when he values the object at a is

$$\sum_{b \in W} p_{ab}[Q^1_{ab}a - T^1_{ab}].$$

A similar expression holds for bidder 2. So, the auctioneer maximizes expected revenue if she can choose Q^j and T^j so that for all $a \in W$ bidder 1's expected profit is zero, i.e.,

$$\sum_{b \in W} p_{ab}[Q^1_{ab}a - T^1_{ab}] = 0,$$

and bidder 2's expected profit for all $b \in W$ is zero, i.e.,

$$\sum_{a \in W} p_{ab}[Q^2_{ab}b - T^2_{ab}] = 0.$$

Substituting this into the incentive compatibility and individual rationality constraints, the auctioneer seeks a solution to:

$$\sum_{b \in W} p_{ab}[Q^1_{kb}a - T^1_{kb}] \leq 0, \quad \forall k \in W \setminus a, \ a \in W,$$

$$\sum_{a \in W} p_{ab}[Q^2_{ak}b - T^2_{ak}] \leq 0, \quad \forall k \in W \setminus b, \ b \in W,$$

$$\sum_{b\in W} p_{ab}[Q^1_{ab}a - T^1_{ab}] = 0, \quad \forall a \in W,$$

$$\sum_{a\in W} p_{ab}[Q^2_{ab}b - T^2_{ab}] = 0, \quad \forall b \in W.$$

Now fix the value of Q^j in the inequalities above and ask if there is a feasible T^j. Rewriting the above inequalities by moving terms that are fixed to the right-hand side (with a change in index on the last two to make the Farkas alternative easier to write out):

$$-\sum_{b\in W} p_{ab}T^1_{kb} \le -\sum_{b\in W} p_{ab}Q^1_{kb}a, \quad \forall k \in W \setminus a,\ a \in W,$$

$$-\sum_{a\in W} p_{ab}T^2_{ak} \le -\sum_{a\in W} p_{ab}Q^2_{ak}b, \quad \forall k \in W \setminus b,\ b \in W,$$

$$\sum_{b\in W} p_{kb}T^1_{kb} = \sum_{b\in W} p_{kb}Q^1_{kb}k, \quad \forall k \in W,$$ (2.1)

$$\sum_{a\in W} p_{ak}T^2_{ak} = \sum_{k\in W} p_{ak}Q^2_{ak}k, \quad \forall k \in W.$$

Let y^1_{ak} be the variable associated with the first inequality, y^2_{kb} be associated with second inequality, z^1_k with the third and z^2_k with the fourth set of inequalities.

Before passing to the Farkas alternative it will be useful to write out the matrix of coefficients associated with the T^1 variables. Assume for this purpose only that $W = \{a, b, c\}$.

$$\begin{bmatrix}
T^1_{aa} & T^1_{ab} & T^1_{ac} & T^1_{bb} & T^1_{ba} & T^1_{bc} & T^1_{cc} & T^1_{ca} & T^1_{cb} \\
0 & 0 & 0 & -p_{ab} & -p_{ab} & -p_{ac} & 0 & 0 & 0 \\
0 & 0 & 0 & 0 & 0 & 0 & -p_{ab} & -p_{aa} & -p_{ac} \\
-p_{ba} & -p_{bb} & -p_{bc} & 0 & 0 & 0 & 0 & 0 & 0 \\
0 & 0 & 0 & 0 & 0 & 0 & -p_{bc} & -p_{ba} & -p_{bb} \\
-p_{ca} & -p_{cb} & -p_{cc} & -p_{cb} & -p_{ca} & -p_{cc} & 0 & 0 & 0 \\
p_{aa} & p_{ab} & p_{ac} & 0 & 0 & 0 & 0 & 0 & 0 \\
0 & 0 & 0 & p_{bb} & p_{ba} & p_{bc} & 0 & 0 & 0 \\
0 & 0 & 0 & 0 & 0 & 0 & p_{cc} & p_{ca} & p_{cb}
\end{bmatrix}$$

Each column of this matrix gives rise to an equation in the alternative. The alternative appears below and the reader may find it helpful to compare it with the columns of the above matrix.

The Farkas lemma asserts that there is no solution to the system (2.1) if there is a solution to the system:

$$-\sum_{a \neq k} p_{ab} y_{ak}^1 + p_{kb} z_k^1 = 0, \quad \forall k, b \in W,$$

$$-\sum_{b \neq k} p_{ab} y_{kb}^2 + p_{ak} z_k^2 = 0, \quad \forall a, k \in W,$$

$$y \geq 0$$

such that

$$-\sum_{a \in W} \sum_{k \neq a} \left[\sum_{b \in W} p_{ab} Q_{kb}^1 a \right] y_{ak}^1 - \sum_{b \in W} \sum_{k \neq b} \left[\sum_{b \in W} p_{ab} Q_{ak}^2 a \right] y_{kb}^2$$

$$+ \sum_{k \in W} \sum_{b \in W} p_{kb} Q_{kb}^1 k z_k^1 + \sum_{k \in W} \sum_{b \in W} p_{kb} Q_{kb}^2 k z_k^2 < 0.$$

Using the first equation, non-negativity of the p's and the y's we conclude that the z's must be non-negative as well. The last inequality which must hold strictly prevents, all of the y variables being zero. Given this, the first equation contradicts the cone assumption made earlier. Thus, the Farkas alternative has no solution, implying that (2.1) has a solution.

Problems

2.1 Sketch the cone generated by the columns of the matrix below:

$$\begin{bmatrix} 2 & 0 & -1 \\ 1 & 1 & 2 \end{bmatrix}$$

What is the cone generated by just the first and third columns of the matrix? If A is the matrix above and $b = (1, 0)$ decide if the system $Ax = b$ has a solution with $x \geq 0$.

2.2 Sketch the cone generated by the columns of the matrix below:

$$\begin{bmatrix} 2 & 1 & -3 \\ -1 & 3 & -2 \end{bmatrix}.$$

2.3 Convert the following system of inequalities/equalities into standard form and then write down its Farkas alternative:

$$x_1 + 2x_2 + 3x_3 \leq 5,$$

$$x_1 + 3x_2 - 2x_3 \geq 7,$$

$$x_1 + x_2 + x_3 \leq 2,$$

$$x_1 - 2x_2 - 3x_3 \geq 3,$$

$$x_1 \qquad\qquad \geq 0.$$

2.4 Use the Farkas lemma to decide if the following system has a non-negative solution:

$$\begin{bmatrix} 4 & 1 & -2 \\ 1 & 0 & 5 \end{bmatrix} \begin{bmatrix} x_1 \\ x_2 \\ x_3 \end{bmatrix} = \begin{bmatrix} -2 \\ 3 \end{bmatrix}.$$

2.5 Let A be an $m \times n$ matrix. Prove, using the Farkas lemma, the following:

The system $Ax \geq b$ has a non-negative solution or there is a non-negative $y \in \mathbb{R}^m$ such that $yA \leq 0$ and $yb > 0$, but not both.

2.6 Let A be an $m \times n$ matrix. Prove, using the Farkas lemma, the following:

The system $Ax = 0$, $\sum_{j=1}^{n} x_j = 1$ has a non-negative solution or there is a $y \in \mathbb{R}^m$ such that $yA \gg 0$, but not both.

2.7 Let A be an $m \times n$ matrix. Prove the following:

The system $Ax = 0$, has a non-zero, non-negative solution or there is a $y \in \mathbb{R}^m$ such that $yA \gg 0$, but not both.

2.8 An $n \times n$ matrix A is called a **Markov matrix** if $a_{ij} \geq 0$ for all i, j and $\sum_{i=1}^{n} a_{ij} = 1$ for all j. These matrices arise in the study of Markov chains and the $\{ij\}^{\text{th}}$ entry is the probability of moving from state j to state i. A vector $x \in \mathbb{R}^n$ is called a **probability vector** if $x_j \geq 0$ for all j and $\sum_{j=1}^{n} x_j = 1$. A probability vector x is called a steady state vector of A if $Ax = x$. Use the Farkas lemma to show that every Markov matrix has a steady state vector.

2.9 Let A be an $m \times n$ matrix. Prove, using the Farkas lemma, the following:

The system $Ax \ll b$ has a solution if and only if $y = 0$ is the only solution for $\{yA = 0, \ yb \leq 0, \ y \geq 0\}$.

2.10 Let A be an $m \times n$ real matrix and $F = \{x \in \mathbb{R}^n : Ax \leq 0\}$. Let $c \in \mathbb{R}^n$ and $G = \{x \in \mathbb{R}^n : cx \leq 0\}$. Use the Farkas lemma to prove that $F \subseteq G$ iff there exists $y \in \mathbb{R}^m_+$ such that $c = yA$.[21]

2.11 Let A be an $m \times n$ matrix and $b \in \mathbb{R}^n$. Use the Farkas lemma to prove that there exist $x \in \mathbb{R}^n$, $y \in \mathbb{R}^m$ such that:

$$Ax \geq 0, \ A^T y = 0, \ y \geq 0, \ a_1 \cdot x + y_1 > 0.$$

Here a_1 denotes the first row of A.

2.12 Suppose A is an $n \times n$ matrix such that $x^T A x = 0$ for all $x \in \mathbb{R}^n$. Show that the system

$$(I + A)x \gg 0, \quad Ax \geq 0, \quad x \geq 0$$

has a solution.

Notes

1 If Mathematics is the Queen of the Sciences, then Carl Friedrich Gauss (1777–1855) is her Prince. A brief and engaging biography can be found in Eric Temple Bell's *Men of Mathematics*. In a choice between accuracy and spice, Bell prefers spice and is delightfully liberal with their application.

   At the age of seven, Gauss is supposed to have summed the integers from 1 to 100 by observing that the sum could be written as the sum of 50 pairs of numbers each pair summing to 101.

2 Guyla Farkas (1847–1930) was a Hungarian Theoretical Physicist. The lemma that bears his name was announced by him in 1894 and has its roots in the problem of specifying the equilibrium of a system. The associated proof was incomplete. A complete proof was published in Hungarian by Farkas in 1898. It is more common to refer to the German version that appeared in the *J. Reine Angew Math.*, 124, 1901, 1–27. The title was *Theorie der Einfachen Ungleichungen*.

3 This is where carefully specifying the indices in steps 1 and 4 matters.

4 The exposition in this section is based in part on Nau and McCardle (1992).

5 It is said that a genius is one who has two great ideas. Frank Plumpton Ramsey (1903–1927) had at least three. Ramsey growth model, Ramsey theory, Ramsey pricing.

6 Ramsey (1931); However he gives no formal proofs.

7 Called the 'incomparable'. De Finetti also introduced the concept of exchangeability.

8 Savage (1954).

9 To hold a long position is to acquire the asset in the hope that its value will increase. To hold a short position is to make a bet that the asset will decline in value. On the stock market this is done by selling a stock one does not own now and buying it back at a later date presumably when its price is lower. In practice, one's broker will 'borrow' the stock from another client and sell it in the usual way. At some time after the sale one tells the broker stop, and buys the stock borrowed at the prevailing price and returns them to the 'lender'. The strategy yields a profit if the price of the stock goes down.

10 The basis set is referred to, for obvious reasons, as a spanning set of securities.

11 The earliest record of a call option is to be found in Aristotle's **Politics**. Six months prior to the olive harvest in spring, the philosopher Thales (62?– 546 B.C.) purchased the right to lease, at the current low rates, oil presses during the harvest. A bumper crop of olives in the spring allowed Thales to sublet the presses at premium rates. He used the resulting profits to support his philosophizing. In his day Thales was remembered more for his philosophy than for his financial acumen. Diogenes Laertius, biographer of the ancient Greek philosophers, urges that all discourse should begin with a reference to Thales.

12 This is called a European option. More elaborate options are available with names like American, Asian and Exotic.

13 The highest bidder was called the *emptor*, from whence the term *caveat emptor*.

14 'It was at Rome, on the 15th October 1764', Gibbon writes, 'as I sat musing amid the ruins of the capitol, while the bare-footed friars were singing vespers in the temple of Jupiter, that the idea of writing the decline and fall of the city first started to my mind'. 71 chapters, 2,136 paragraphs, a million and a half words, 8,000 footnotes and one American revolution later, Gibbon produced *The Decline and Fall of the Roman Empire*. The incident of the auction is described in chapter V, volume I.

15 Bodyguard of the Emperor.

16 Pertinax himself had secured the empire by promising huge bribes to the Praetorians. Upon taking the purple, he reneged and the guard took their revenge.

17 The description by Gibbons is worth a read: 'This infamous offer, the most insolent excess of military license, diffused an universal grief, shame, and indignation throughout

the city. It reached at length the ears of Didius Julianus, a wealthy senator, who, regardless of the public calamities, was indulging himself in the luxury of the table. His wife and his daughter, his freedmen and his parasites, easily convinced him that he deserved the throne, and earnestly conjured him to embrace so fortunate an opportunity. The vain old man hastened to the Praetorian camp, where Suplicianus was still in treaty with the guards, and began to bid against him from the foot of the rampart'.

18 This section is based on Cremer and McLean (1988).

19 In other books and courses you will learn about the revelation principle which explains why this formulation is general enough to encompass all auctions.

20 This is known as the **common prior** assumption.

21 This was the manner in which Farkas originally stated his lemma.

References

Aristotle: 1999, *Politics*, Clarendon Aristotle series, Clarendon Press, New York.

Black, F. and Scholes, M. S.: 1973, The pricing of options and corporate liabilities, *Journal of Political Economy* **81**(3), 637–54.

Cremer, J. and McLean, R. P.: 1988, Full extraction of the surplus in bayesian and dominant strategy auctions, *Econometrica* **56**(6), 1247–57.

Finetti, B. D.: 1937, La prevision: Ses lois logiques, ses sources subjectives, *Ann. Inst. Henri Poincare* **7**, 1–68.

Gibbon, E. and Bury, J. B.: 1896, *The history of the decline and fall of the Roman empire*, Macmillan & Co., New York.

Merton, R. C.: 1973, Theory of rational option pricing, *The Bell Journal of Economics and Management Science* **4**(1), 141–83.

Nau, R. F. and McCardle, K. F.: 1992, Arbitrage, rationality, and equilibrium, *in* J. Geweke (ed.), *Decision making under risk and uncertainty: New models and empirical findings*, Theory and Decision Library, Mathematical and Statistical Methods, vol. 22. Norwell, Mass. and Dordrecht, Kluwer Academic, pp. 189–99.

Ramsey, F. P. and Braithwaite, R. B.: 1931, *The foundations of mathematics and other logical essays*, International library of psychology, philosophy, and scientific method, Harcourt K. Paul, Trench, Trubner & Co. Ltd., New York.

Savage, L. J.: 1954, *The foundations of statistics*, Wiley publications in statistics, Wiley, New York.

Varian, H. R.: 1987, The arbitrage principle in financial economics, *Journal of Economic Perspectives* **1**(2), 55–72.

3 Convex sets

Definition 3.1 *A set C of vectors/points is called* **convex** *if for all $x, y \in C$ and $\lambda \in [0, 1]$, $\lambda x + (1 - \lambda)y \in C$.*

Geometrically, a set is convex if any two points within it can be joined by a straight line that lies entirely within the set. Equivalently, the weighted average of any two points in C is also in C. The quintessential convex set is the region enclosed by a circle. Figure 3.1 shows two sets. The one on the left is convex while the one on the right is not. One implication of convexity is the following: if $x^1, x^2, \ldots, x^r$ are a finite collection of vectors in a convex set C, then $\sum_{i=1}^{r} \lambda_i x^i$ is also in C where $\sum_{i=1}^{r} \lambda_i = 1$ and $\lambda_i \geq 0$ for all i. One could just as well define convexity of a set C by requiring that the weighted average of any finite subset of points in C also be in C. Verifying convexity would require checking this condition for every finite subset of points. The definition given above says that it suffices to check every pair of points, presumably a less laborious task.

Convex sets have many useful properties. The easiest ones to establish are summarized below without proof:

1. The set $\{x: Ax = b, \ x \geq 0\}$ is convex.
2. If C is convex then $\alpha C = \{y: y = \alpha x, \ x \in C\}$ is convex for all real α.
3. If C and D are convex sets, then the set $C + D = \{y: y = x + z, \ x \in C, \ z \in D\}$ is convex.
4. The intersection of any collection of convex sets is convex.

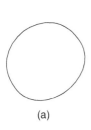   

(a)   (b)

Figure 3.1

3.1   Separating hyperplane theorem

One of the most important results about convex sets is the separating hyperplane theorem. Given a point x and a convex set C not containing x, one should be able to draw a straight line that separates the two, i.e., the point is on one side and the set C on the other side. Figure 3.2 illustrates the theorem. The requirement that the line be straight is what makes it non-trivial. To see why such a statement should be true, consider Figure 3.3. A portion of the border of our convex set C is shown, with a dotted line. The point b, conveniently chosen to be at the origin is not in the set C. The point x^* is the point in C closest to b, conveniently chosen to be on the horizontal axes. The figure assumes that $b \neq x^*$. The line labeled L is perpendicular to the segment $[b, x^*]$ and is chosen to be midway between x^* and b. The line L is our candidate for the straight line that separates b from C. For the line L to be our separator, we need to show that no point $y \in C$ lies to the left of L. Suppose not as shown in Figure 3.4. Since $y \in C$, by convexity of

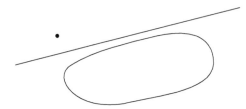

Figure 3.2

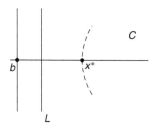

Figure 3.3

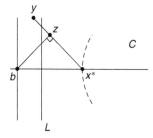

Figure 3.4

C, every point on the line segment joining y to x^* is also in C. In particular, the point z marked on Figure 3.4 is in C. The point z is chosen so that the line joining b to z is perpendicular to the line joining x^* to y. Clearly z is closer to b than x^*, contradicting the choice of x^*.

Lemma 3.2 *Let C be a compact set not containing the origin. Then there is an $x^0 \in C$ such that $d(x^0, 0) = \inf_{x \in C} d(x, 0) > 0$.*

Proof Follows from the continuity of the distance function and the Weierstrass theorem. ∎

Definition 3.3 *A **hyperplane** $H = (h, \beta)$ where $h \in \mathbb{R}^n$ and $\beta \in \mathbb{R}$ is the set $\{x \in \mathbb{R}^n : hx = \beta\}$. A **half-space** is the set $\{x \in \mathbb{R}^n : hx \leq \beta\}$. The set of solutions to a single equation form a hyperplane. The set of solutions to a single inequality form a half-space.*

Figure 3.5(a) illustrates a hyperplane in $\mathbb{R}^2$ while Figure 3.5(b) illustrates a half-space.

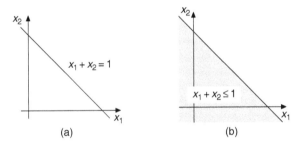

(a)   (b)

Figure 3.5

It is easy to see that a hyperplane and a half-space are both convex sets.

Theorem 3.4 (Strict separating hyperplane theorem) *Let C be a closed convex set and $b \notin C$. Then there is a hyperplane (h, β) such that $hb < \beta < hx$ for all $x \in C$.*

Proof By a translation of the coordinates we may assume that $b = 0$. Choose $x^0 \in C$ that minimizes $d(x, 0)$ for $x \in C$. By Lemma 3.2, such an x^0 exists and $d(x^0, 0) > 0$. The reader will note that Lemma 3.2 assumes compactness but here we do not. Here is why. Pick any $y \in C$ and let $C' = C \cap \{x \in C : d(x, 0) \leq d(y, 0)\}$. Notice that C' is the intersection of two closed sets and so is closed as well. It is also bounded. It is easy to see that the point in C' closest to 0 is also the point in C closest to 0.

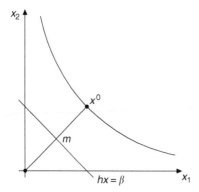

Figure 3.6

Let m be the midpoint of the line joining 0 to x^0, i.e., $m = x^0/2$. Choose (h, β) to be the hyperplane that goes through m and is perpendicular to the line joining 0 to x^0 (see Figure 3.6). Formally we choose h to be the vector x^0 scaled by $d(x^0, 0)$, i.e., $h = x^0/d(x^0, 0)$. Set $\beta = h \cdot m$. Notice $hm = x^0/2 \cdot x^0/d(x^0, 0) = d(x^0, 0)/2$.

Next, we verify that $b = 0$ is on one side of the hyperplane (h, β) and x^0 is on the other. Observe that $hb = 0 < d(x^0, 0)/2 = hm$. Next,

$$hx^0 = x^0 \frac{x^0}{d(x^0, 0)} = d(x^0, 0) > \frac{d(x^0, 0)}{2} = hm.$$

Now we show that all $x \in C$ are on the same side of (h, β) as x^0, that is $h \cdot x > \beta$ for all $x \in C$. Pick any $x \in C$ different from x^0. By the convexity of C, $(1 - \lambda)x^0 + \lambda x \in C$. From the choice of x^0, $d(x^0, 0)^2 \leq d((1 - \lambda)x^0 + \lambda x, 0)^2$. Since $d(z, 0)^2 = z \cdot z$ we have

$$d(x^0, 0)^2 \leq ((1 - \lambda)x^0 + \lambda x, 0)) \cdot ((1 - \lambda)x^0 + \lambda x, 0))$$
$$\leq d(x^0, 0)^2 + 2\lambda x^0(x - x^0) + \lambda^2 d(x - x^0, 0)^2$$

which reduces to

$$0 \leq 2x^0(x - x^0) + \lambda d(x - x^0, 0)^2.$$

Since λ can be made arbitrarily small it follows that $x^0 \cdot (x - x^0) \geq 0$ for all $x \in C$. Using the fact that $h = x^0/d(x^0, 0)$ and $x^0 = 2m$, we can rewrite this last inequality as

$$0 \leq [d(x^0, 0)h](x - 2m),$$

i.e., $hx \geq 2mh > hm$. ∎

The conclusion of the theorem is usually phrased as follows: a hyperplane H strictly separates C from b. If one drops the requirement that C be closed, one obtains a weaker conclusion.

Theorem 3.5 (Weak separating hyperplane theorem) *Let C be a convex set and $b \notin C$. Then there is a hyperplane (h, β) such that $hb \leq \beta \leq hx$ for all $x \in C$.*

Proof The proof is similar to the previous one. The only difference is in the choice of the point x^0. It is chosen so that $d(x^0, 0) = \inf_{x \in C} d(x, 0)$. Since it is possible that $x^0 = b$ (e.g., if b were on the boundary of C), the strict inequalities in the previous theorem must be replaced by weak inequalities. ∎

Theorem 3.6 *Let C and D be two non-empty, disjoint, convex sets in $\mathbb{R}^n$. Then there is a hyperplane (h, β) such that $hx \geq \beta \geq hy$ for all $x \in C$ and $y \in D$.*

Proof Let $K = \{z : z = x - y, \, x \in C, \, y \in D\}$, the set of vectors that can be expressed as a difference between a vector in C and one in D. The set K is convex and since C and D are disjoint, does not contain the origin. By the weak separating hyperplane theorem there is a hyperplane (h, β') such that $h \cdot 0 \leq \beta \leq h \cdot z$ for all $z \in K$. Pick any $x \in C$ and $y \in D$ then $x - y \in K$. Therefore $h \cdot x - h \cdot y \geq 0$ for all $x \in C$ and $y \in D$. In particular:

$$h \cdot x \geq \inf_{u \in C} h \cdot u \geq \sup_{v \in D} h \cdot v \geq h \cdot y.$$

Choose $\beta \in [\inf_{u \in C} h \cdot u, \sup_{v \in D} h \cdot v]$ to complete the proof. ∎

Example 9 *It is natural to conjecture that Theorem 3.6 can be strengthened to provide strict separation if we assume C and D to be closed. This is false. Let $C = \{(x_1, x_2) : x_1 \geq 0, \, x_2 \geq 1/x_1\}$ and $D = \{(x_1, x_2) : x_2 = 0\}$. Both sets are closed, convex and disjoint, see Figure 3.7. However, strict separation is not possible. To see why not, observe that for all positive numbers n, $(n, 1/n) \in C$ while $(n, 0) \in D$. As $n \to \infty$ the point $(n, 1/n)$ approaches $(n, 0)$.*

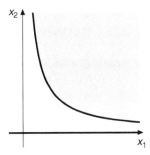

Figure 3.7

Theorem 3.7 *Let C and D be two non-empty, disjoint, closed, convex sets in $\mathbb{R}^n$ with C being bounded. Then there is a hyperplane (h, β) such that $hx > \beta > hy$ for all $x \in C$ and $y \in D$.*

Proof   The proof is similar to the proof of the previous theorem. Let $K = \{z : z = x - y, x \in C, y \in D\}$. The set K is clearly convex. If K is closed we can apply the strict separating hyperplane theorem to obtain the result.

It remains then to prove that K is closed. Let $\{z^n\}_{n \geq 1}$ be a convergent sequence in K with limit z^* which may or may not be in K. For each n there is a $x^n \in C$, $y^n \in D$ such that $z^n = x^n - y^n$. By the Bolzano–Weierstrass theorem the sequence $\{x^n\}_{n \geq 1}$ has a convergent subsequence with limit $x^* \in C$, say. Since $y^n = x^n - z^n \to x^* - y^*$, $\{y^n\}_{n \geq 1}$ has a limit, call it $y^* \in D$. Thus z^* is the difference between a vector in C and one in D, i.e., $z^* \in K$.   ∎

The strict separating hyperplane yields the Farkas lemma as a special case as well as a geometric interpretation of the same. To apply it we need two results.

Lemma 3.8 *Let A be an $m \times n$ matrix, then $\text{cone}(A)$ is a convex set.*

Proof   Pick any two $y, y' \in \text{cone}(A)$. Then there exist $x, x' \geq 0$ such that

$$y = Ax, \quad y' = Ax'.$$

For any $\lambda \in [0, 1]$ we have

$$\lambda y + (1 - \lambda)y' = \lambda Ax + (1 - \lambda)Ax' = A(\lambda x + (1 - \lambda)x').$$

Since $\lambda x + (1 - \lambda)x' \geq 0$ it follows that $\lambda y + (1 - \lambda)y' \in \text{cone}(A)$.   ∎

The proof of the next lemma introduces an argument that will be used again later.

Lemma 3.9 *Let A be an $m \times n$ matrix, then $\text{cone}(A)$ is a closed set.*

Proof   First suppose that B is a matrix all of whose columns form a LI set. We prove that $\text{cone}(B)$ is closed. Let $\{w^n\}$ be any convergent sequence in $\text{cone}(B)$ with limit w. We must show that $w \in \text{cone}(B)$. For each w^n there is a $x^n \geq 0$ such that $w^n = Bx^n$. We use the fact that Bw^n converges to show that x^n converges.

Since $B^T B$ and B have equal rank, $B^T B$ is invertible and $x^n = (B^T B)^{-1} B^T (Bx^n)$. Hence, if $Bx^n \to w$, $x^n \to (B^T B)^{-1} B^T w$. Therefore $\text{cone}(B)$ is closed.[1]

Now we show that $\text{cone}(A)$ is closed. Let B be any LI subset of columns of A. By the above $\text{cone}(B)$ is closed. The union of all $\text{cone}(B)$'s where B is a LI subset of columns is $\text{cone}(A)$ because every element of $\text{cone}(A)$ can be expressed as a non-negative linear combination of some LI columns of A.

To see why this last statement is true let $b \in \text{cone}(A)$. Then there is a vector $\lambda \geq 0$ such that $b = A\lambda$. Let $S = \{j : \lambda_j > 0\}$ and B the submatrix of A consisting of the columns in S. Since there may be many ways to express b as a non-negative linear combination of columns of A, we choose an expression that uses the fewest columns. Thus b cannot be expressed as a non-negative linear combination of $|S| - 1$ or fewer columns of A.

If the columns in S are LI we are done. So, suppose not. Since the columns of B are LD, $\ker(B) \neq \{0\}$. Since $\ker(B) \neq \{0\}$ we can choose a non-zero $\mu \in \ker(B)$. Consider $\lambda - t\mu$ for any $t \in \mathbb{R}$. Notice that $b = A(\lambda - t\mu)$.

Choose t so that $\lambda_j - t\mu_j \geq 0$ for all $j \in S$ and $\lambda_j - t\mu_j = 0$ for at least one $j \in S$. If such a choice of t exists, it would imply that b is a non-negative linear combination of $|S| - 1$ columns of B, a contradiction which proves the claim.

To see that such a t exists suppose first that $\mu_j > 0$ for all $j \in S$. Then, set $t = \min_{j \in S}\{\lambda_j / \mu_j\}$. If at least one of $\mu_j < 0$, set $t = -\max_{\mu_j < 0}\{|\lambda_j / \mu_j|\}$. For any j such that $\mu_j > 0$, we have that $\lambda_j - t\mu_j > 0$ since $t < 0$ and μ_j, $\lambda_j \geq 0$. For any j such that $\mu_j < 0$ we have that

$$\lambda_j - t\mu_j \geq \lambda_j + \left|\frac{\lambda_j}{\mu_j}\right|\mu_j = 0$$

by the choice of t.

Now, there are a finite number of such B's. The union of a finite number of closed sets is closed, so $\text{cone}(A)$ is closed. ∎

Theorem 3.10 (Farkas lemma) *Let A be an $m \times n$ matrix, $b \in \mathbb{R}^m$ and $F = \{x \in \mathbb{R}^n : Ax = b, x \geq 0\}$. Then either $F \neq \emptyset$ or there exists $y \in \mathbb{R}^m$ such that $yA \geq 0$ and $yb < 0$ but not both.*

Proof The 'not both' part of the result is obvious. Now suppose $F = \emptyset$. Then $b \notin \text{cone}(A)$. Since $\text{cone}(A)$ is convex and closed we can invoke the strict separating hyperplane theorem to identify a hyperplane, (h, β) that separates b from $\text{cone}(A)$. Without loss of generality we can suppose that $h \cdot b < \beta$ and $h \cdot z > \beta$ for all $z \in \text{cone}(A)$. Since the origin is in $\text{cone}(A)$ it is easy to see that $\beta < 0$.

Let a^j be the jth column vector of the matrix A. We show that $h \cdot a^j \geq 0$. Suppose not, i.e., $h \cdot a^j < 0$. Notice that $\lambda a^j \in \text{cone}(A)$ for any $\lambda \geq 0$. Thus

$$h \cdot [\lambda a^j] > \beta.$$

Since $\lambda > 0$ can be chosen arbitrarily large, $h \cdot [\lambda a^j]$ can be made smaller than β, a contradiction. Thus $h \cdot a^j \geq 0$ for all columns j. Hence, $y = h$ is our required vector. ∎

Unlike our earlier proof of the Farkas lemma we cannot conclude from the separating hyperplane theorem that when $b \notin \text{cone}(A)$ and A has rank r that there are r LI columns of A, $\{a^1, a^2, \ldots, a^{r-1}\}$ and a $y \in \mathbb{R}^m$ such that $y \cdot a^j = 0$ for $1 \leq j \leq r - 1$, $y \cdot a^j \geq 0$ for all $j \geq r$ and $y \cdot b < 0$.

3.2   Polyhedrons and polytopes

This section shows how certain kinds of convex sets can be represented as the intersection of half spaces or as a weighted average of a finite number of points.

Recall that a set C of vectors is called a **cone** if $\lambda x \in C$ whenever $x \in C$ and $\lambda > 0$.

Definition 3.11   *A cone $C \subset \mathbb{R}^n$ is **polyhedral** if there is a matrix A such that $C = \{x \in \mathbb{R}^n : Ax \leq 0\}$.*

Geometrically, a polyhedral cone is the intersection of a finite number of half-spaces through the origin.

Example 10   *Figure 3.8 illustrates a polyhedral cone in $\mathbb{R}^2$. The polyhedral cone is the darker of the regions.*

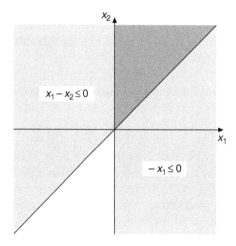

Figure 3.8

Not every cone is polyhedral. Consider the set in $\mathbb{R}^2$ that is the union of all vectors of the form $(0, z)$ and $(z, 0)$ where $z \geq 0$. This is a cone but is clearly not polyhedral. In fact it is not even convex.

Returning to Figure 3.8, we see that the polyhedral cone can also be expressed as the cone generated by the vectors $(0, 1)$ and $(1, 1)$. This is no coincidence. Given a polyhedral cone, identify the 'bounding' hyperplanes. The cone generated by the normal's to these hyperplanes will coincide with the initial polyhedral cone. In Figure 3.8, the hyperplane $-x_1 = 0$ written in dot product form is $(-1, 0) \cdot (x_1, x_2) = 0$. The normal to this hyperplane is $(0, 1)$. The second hyperplane is $(1, -1) \cdot (x_1, x_2) = 0$ and the corresponding normal is $(1, 1)$.

The converse is also true. Consider a matrix A and the cone, cone(A), generated by its columns. Then cone(A) will be polyhedral. To see why this is plausible, we deduce from the Farkas lemma that $b \in$ cone(A) iff $y \cdot b \geq 0$ for all $yA \geq 0$. Let $F = \{y: yA \geq 0\}$, then a vector a is in cone(A) iff $y \cdot a \geq 0$ for every $y \in F$. In other words,

$$\text{cone}(A) = \{a: (-y) \cdot a \leq 0 \; \forall y \in F\}.$$

Thus cone(A) can be expressed as the intersection of a collection of half spaces of the form $y \cdot a \leq 0$. The next theorem sharpens this conclusion by showing that cone(A) can be expressed as the intersection of a *finite* number of half spaces. This establishes that cone(A) is polyhedral.

Theorem 3.12 (Farkas–Minkowski–Weyl) [2]   *A cone C is polyhedral iff there is a finite matrix A such that $C = $ cone(A).*

Proof   Let A be a $m \times n$ matrix with rank r. Let $C = $ cone(A). We prove that C is polyhedral.

Suppose first that $m = r$. For each LI subset S of $r - 1$ columns of A we can find a non-trivial vector z^S such that $z^S \cdot a^j = 0$ for all $j \in S$. The system

$$z^S \cdot a^j = 0 \quad \forall j \in S$$

consists of r variables and $r - 1$ LI equations. So, the set of solutions forms a one-dimensional subspace. In particular every solution can be expressed as a scalar multiple of just one solution, y^S, say. Since there are only a finite number of choices for S, there are, over all possible choices of LI subsets of columns of A, a finite number of these y^S vectors.

Consider any $b \notin$ cone(A). By the Farkas lemma, there is an r-set S of LI columns of A and a vector y^b such that $y^b \cdot a^j = 0$ for all $j \in S$ and $y^b \cdot a^j \geq 0$ for all $j \notin S$. Hence y^b must be one of the y^S vectors identified above. Thus the set $F = \cup_{b \notin \text{cone}(A)} y^b$ is finite.

Any $y \in F$ has the property that $y \cdot a^j \geq 0$ for all columns a^j of A. Hence $y \cdot x \geq 0$ for all $x \in$ cone(A) because each x is a non-negative linear combination of columns of A. So, cone(A) $\subseteq \cap_{y \in F} \{y \cdot x \geq 0\}$. If $x \notin$ cone(A), there is a $y \in F$ such that $y \cdot x < 0$. Hence cone(A) $= \cap_{y \in F} \{y \cdot x \geq 0\}$.

Now suppose $m < r$. Without loss of generality we may assume that the first r rows of A are LI. Let A' be the matrix of the first r rows of A and b' the vector of the first r components of b. By the previous argument we know that cone(A') is polyhedral. Let F' be the (finite) set of half-spaces in $\mathbb{R}^r$ whose intersection defines cone(A'). We can extend any $y \in F'$ into a vector in $\mathbb{R}^m$ by adding $m - r$ components all of value zero.

Consider any $b \notin \text{cone}(A)$. Suppose first that $b \notin \text{span}(A)$. Then there is a $y^b \in \mathbb{R}^m$ such that $y^b A = 0$ but $y^b \cdot b < 0$. The space of solutions to $yA = 0$ has dimension $m - r$. So, we can always choose y^b to be one of the bases vectors of this space. Thus the set $F = \cup_{b \notin \text{span}(A)} y^b$ is finite. Observe that for all $y \in F$, $y \cdot x = 0$ for all $x \in \text{cone}(A)$ and $y \cdot b < 0 \; \forall b \notin \text{span}(A)$.

We now show that $\text{cone}(A) = \cap_{y \in F' \cup F} \{x: yx \geq 0\}$. Clearly $\text{cone}(A) \subseteq \cap_{y \in F' \cup F} \{x: yx \geq 0\}$. Consider a $b \notin \text{cone}(A)$. If $b \notin \text{span}(A)$, then there is a $y \in F$ such that $y \cdot b > 0$, i.e., $b \notin \cap_{y \in F' \cup F} \{x: yx \geq 0\}$. If $b \in \text{span}(A)$ but $b \notin \text{cone}(A)$, then $b' \notin \text{cone}(A')$. Again, $b \notin \cap_{y \in F' \cup F} \{x : yx \geq 0\}$.

For the other direction suppose C is a polyhedral cone, i.e. $C = \{x \in \mathbb{R}^n: Ax \leq 0\}$ where A is an $m \times n$ matrix. By the previous argument the cone generated by the *rows* of A, $\text{cone}(A^T)$, is polyhedral, i.e. $\text{cone}(A^T) = \{y \in \mathbb{R}^m: B^T y \leq 0\}$ for some matrix B. Now the rows of B^T (equivalently the columns of B) are in C. To see why, observe that $b^j \cdot a^k \leq 0$ for any column j of B and row a^k of A. Thus $\text{cone}(B) \subseteq C$. Suppose there is a $z \in C \backslash \text{cone}(B)$. Since $\text{cone}(B)$ is polyhedral, there is a vector w such that $w \cdot b^j \leq 0$ for all j and $w \cdot z > 0$. But this implies that $w \in \text{cone}(A^T)$, i.e., $w \cdot x \leq 0$ for all $x \in C$ a contradiction. ∎

Definition 3.13 *A non-empty set $P \subset \mathbb{R}^n$ is called a **polyhedron** if there is an $m \times n$ matrix A and vector $b \in \mathbb{R}^m$ such that $P = \{x \in \mathbb{R}^n: Ax \leq b\}$.*

Thus a polyhedron is the intersection of finitely many half spaces.

Example 11   *Figure 3.9 illustrates a polyhedron in $\mathbb{R}^2$.*

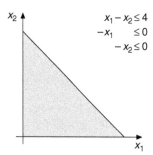

$$x_1 - x_2 \leq 4$$
$$-x_1 \quad \leq 0$$
$$- x_2 \leq 0$$

Figure 3.9

A set of the form $\{x \in \mathbb{R}^n: Ax \leq b, \; A'x = b'\}$, in spite of the equality constraints, is also called a polyhedron. To see why this is legitimate consider the set $P = \{(x_1, x_2, x_3) \in \mathbb{R}^3: x_1 + x_2 + x_3 \leq 3, \; x_3 = 1\}$. We can use the equality constraint $x_3 = 1$, to eliminate the variable x_3 from the system to yield $Q = \{(x_1, x_2) \in \mathbb{R}^2: x_1 + x_2 \leq 2\}$, which is a polyhedron in $\mathbb{R}^2$. Every point in P corresponds to a point in Q and vice-versa. Because of this correspondence we can interpret P to be a polyhedron but one 'living' in a lower dimensional space. In general, the set of feasible solutions to a system of inequalities and equations in $\mathbb{R}^n$ can always be interpreted to be a polyhedron in a lower dimensional space. The inequality representation in lower dimensions is obtained by using the equality constraints to eliminate some of the variables.

Definition 3.14 *Let $S \subset \mathbb{R}^n$. A vector x can be expressed as a* **convex combination** *of vectors in S if there is a finite set $\{v^1, v^2, \ldots, v^m\} \subset S$ such that $x = \sum_{j=1}^{m} \lambda_j v^j$ where $\sum_{j=1}^{m} \lambda_j = 1$ and $\lambda_j \geq 0 \; \forall j$.*

Definition 3.15 *Let $S \subset \mathbb{R}^n$. The* **convex hull** *of S, conv(S), is the set of all vectors that can be expressed as a convex combination of vectors in S.*

Definition 3.16 *A set $P \subset \mathbb{R}^n$ is a called a* **polytope** *if there is a finite set $S \subset \mathbb{R}^n$ such that $P = \text{conv}(S)$.*

Example 12 *The polyhedron of Figure 3.9 is the convex hull of $(0, 0)$, $(0, 1)$ and $(1, 0)$ and so is a polytope. Not every polyhedron is a polytope as Figure 3.10 shows.*

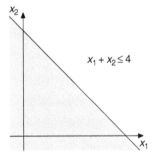

Figure 3.10

Both the Farkas–Minkowski–Weyl theorem and the next result are analogs of the bases theorem of linear algebra. The set of all linear combinations of a finite set of vectors is a vector subspace and, every (finite dimensional) vector subspace can be described as the set of all linear combinations of a finite set of vectors (the bases).

The Farkas–Minkowski–Weyl theorem says that the set of all *non-negative* linear combination of a finite number of vectors is a polyhedral cone. Further, every polyhedral cone can be described as the set of all *non-negative* linear combination of a finite number of vectors.

The next result says that every *convex* combination of a finite number of vectors is a polyhedron. Further, every polyhedron (provided it is bounded) can be expressed as the convex hull of a finite set of vectors.

Theorem 3.17 (Resolution theorem) *A non-empty $P \subset \mathbb{R}^n$ is a polyhedron iff $P = Q + C$, where Q is a polytope and C is a polyhedral cone.*

Proof   Let $P = \{x \in \mathbb{R}^n : Ax \leq b\}$ be a polyhedron and consider the polyhedral cone $\{(x, u) : x \in \mathbb{R}^n, \ u \geq 0, \ Ax - ub \leq 0\}$. It can, by the previous theorem, be generated by finitely many vectors of the form $\{(x^k, u^k)\}_{k \geq 1}$. By normalizing we can assume that $u^k = 0, 1$ for all k. Let Q be the convex hull of vectors of the form $(x^k, 0)$ and C be the cone generated by the vectors of the form $(x^k, 1)$. It is easy to see that $P = Q + C$.

Now suppose $P = Q + C$ where $Q = \text{conv}(\{x^1, x^2, \ldots, x^m\})$ and $C = \text{cone}(\{y^1, y^2, \ldots, y^t\})$. Then $x \in P$ iff $(x, 1)$ is in the cone generated by $\{(x^1, 1), \ldots, (x^m, 1), (y^1, 0), \ldots, (y^t, 0)\}$. By the previous theorem such a cone is polyhedral, i.e., it is equal to $\{(x, u) : x \in \mathbb{R}^n, \ u \geq 0, \ Ax - ub \leq 0\}$ for suitable A and b. Hence $x \in P$ iff $Ax \leq b$.   ∎

Example 13   *Let $P = \{(x_1, x_2) : x_1 + x_2 \leq 4\}$. Choose $Q = \{(2, 2)\}$ and C would be the finite cone generated by $(-1, 1), (1, -1)$ and $(-1, -1)$. Any element of $Q + C$ will have the form*

$$(2, 2) + \lambda_1(-1, 1) + \lambda_2(1, -1) + \lambda_3(-1, -1)$$
$$= (2 - \lambda_1 + \lambda_2 - \lambda_3, 2 + \lambda_1 - \lambda_2 - \lambda_3).$$

To verify that this element is in P we add the two components

$$2 - \lambda_1 + \lambda_2 - \lambda_3 + 2 + \lambda_1 - \lambda_2 - \lambda_3 = 4 - 2\lambda_3 \leq 4.$$

This establishes that $Q + C \subseteq P$. We leave it to the reader to verify that $P \subseteq Q + C$. The example is instructive because it shows that the decomposition implied by the Resolution theorem need not be unique. We could have chosen $Q = \{1, 3\}$ for example.

We will say that P is generated by the vectors $\{x^1, x^2, \ldots, x^m\}$ and the directions $\{y^1, y^2, \ldots, y^t\}$ if

$$P = \text{conv}(\{x^1, x^2, \ldots, x^m\}) + \text{cone}(\{y^1, y^2, \ldots, y^t\}).$$

It is easy to see that P is a polytope iff P is a bounded polyhedron.

Definition 3.18   *Let $S \subset \mathbb{R}^n$ be convex. An **extreme point** of S is a point that cannot be expressed as a convex combination of any points in S.*

Definition 3.19   *Let $S \subset \mathbb{R}^n$ be convex. A **ray** of S is a vector r such that $x + \lambda r \in S$ for all $x \in S$ and $\lambda \geq 0$. An **extreme ray** is one that cannot be expressed as a non-negative linear combination of other rays.*

If P has any extreme points, then they must be contained in the polytope Q identified by the resolution theorem. In fact, in this case we can take Q to be the convex hull of extreme points of P and C to be the set of extreme rays of P. If P has no extreme points then the extreme points of Q will not be extreme points of P. Consider the polyhedron from example 4, $P = \{(x_1, x_2): x_1 + x_2 \leq 4\}$. This is a polyhedron *without* extreme points. Nevertheless, $P = Q + C$ where $Q \neq \varnothing$. In other words, in the decomposition of P into Q and C, the extreme points of Q need not be extreme points of P.

If P is a polytope, then in the resolution theorem decomposition, $C = \varnothing$. In this case P can be described as the convex combination of its extreme points or as the intersection of a finite number of half spaces.

Theorem 3.20 (Caratheodory theorem)[3]   *Let $S \subset \mathbb{R}^n$. Then every $x \in \text{conv}(S)$ can be expressed as a convex combination of at most $n + 1$ points in S.*

Proof   Suppose that $x = \sum_{j=1}^{m} \lambda_j x^j$ where $m \geq n + 2$, $\{x^j\}_{j \geq 1} \in S$, $\sum_{j=1}^{m} \lambda_j = 1$ and $\lambda_j \geq 0$ for all j. It suffices to show that x can be written as a convex combination of $m - 1$ points in S. We may suppose that $\lambda_j > 0$ for all j, otherwise we are done.

Let A be a matrix whose columns are the set $\{x^j\}_{j \geq 1} \in S$ and e the n-vector all of whose components are equal to 1. Then

$$\begin{bmatrix} A \\ e \end{bmatrix} [\lambda] = \begin{bmatrix} x \\ 1 \end{bmatrix}.$$

Since $m \geq n + 2$, the columns of $[\begin{smallmatrix} A \\ e \end{smallmatrix}]$ are LD. Thus $\ker([\begin{smallmatrix} A \\ e \end{smallmatrix}]) \neq \varnothing$. Choose any non-zero $r \in \ker([\begin{smallmatrix} A \\ e \end{smallmatrix}])$ and consider $\lambda - \theta r$ where $\theta \in \mathbb{R}$ will be chosen later. Then

$$x = A(\lambda - \theta r).$$

If we can choose θ so that $\lambda_j - \theta r_j \geq 0 \; \forall j$, $\lambda_j - \theta r_j = 0$ for at least one j and $\sum_{j \in S} \lambda_j - \theta r_j = 1$ we are done. For then we have expressed x as a convex combination of $m - 1$ vectors in S. Repeating this argument completes the proof.

It remains then to show that such a θ can be chosen. Choose θ so that $1/\theta = \max_i r_i/\lambda_i$ and let k be the index where this maximum is achieved. Since $\sum_{j=1}^{m} r_j = 0$, at least one $r_i > 0$ and so $\theta > 0$. Set $q_i = \lambda_i - \theta r_i \geq 0$. Notice that $q_k = 0$ and $\sum_{k=1}^{m} q_i = 1$. However,

$$x = \sum_{j=1}^{m} \lambda_j x^j = \sum_{j=1}^{m} q_j x^j + \theta \sum_{j=1}^{m} r_j x^j = \sum_{j \neq k} q_j x^j. \qquad \blacksquare$$

A consequence of the Caratheodory theorem is that the convex hull of a compact set is also compact.

3.3   Dimension of a set

To motivate the definition of dimension of a set, consider the the polyhedron $P = \{(x_1, x_2) \in \mathbb{R}^2 : x_1 + x_2 = 2\}$. While P sits in a two dimensional space, a sketch of P reveals that it is a straight line, the quintessential one dimensional object. Algebraically, each element of P can be described by a single number. Once one has specified x_1, x_2 is determined. The definition of dimension reconciles the idea of a low dimensional object living in a higher dimensional space.

Definition 3.21   *The collection* $\{x^1, x^2, \ldots, x^k\} \subset \mathbb{R}^n$ *is* **affinely independent** *if the collection* $\{x^2 - x^1, x^3 - x^1, \ldots, x^k - x^1\}$ *is LI.*

The cardinality of the largest set of affinely independent vectors in $\mathbb{R}^n$ is $n + 1$. Take n LI basis vectors and the zero vector.

Definition 3.22   *Let* $S \subset \mathbb{R}^n$ *and suppose the cardinality of the largest set of affinely independent vectors in S is $k + 1$. Then, the* **dimension** *of S, dim(S), is k.*

In other words, the dimension of a set S is the smallest dimensional subspace that contains the set. A set $S \subset \mathbb{R}^n$ is called **full dimensional** if $\dim(S) = n$.

Example 14   *Let* $P = \{(x_1, x_2) \in \mathbb{R}^2 : x_1 + x_2 = 2\}$. *Consider the following set of vectors:* $\{(1, 1), (2, 0)\} \subset P$. *This is clearly affinely independent. So,* $\dim(P) \geq 1$. *To show that* $\dim(P) = 1$, *it suffices to show that P cannot contain a set of three affinely independent vectors.*

To see why, suppose not. Let y^1, y^2, y^3 *be three affinely independent vectors in P. Then* $y^2 - y^1, y^3 - y^1$ *are LI. We know that*

$$(y_1^2 - y_1^1) + (y_2^2 - y_2^1) = 0,$$
$$(y_1^3 - y_1^1) + (y_2^3 - y_3^1) = 0.$$

So, the vectors $y^2 - y^1$, $y^3 - y^1$ *are of the form* $(a, -a)$ *and* $(b, -b)$ *where* $a = y_1^2 - y_1^1$ *and* $b = y_1^3 - y_1^1$. *If the pair* $(a, -a)$ *and* $(b, -b)$ *are identical we are done. If not, one of them is non zero. Suppose then* $a \neq 0$. *Notice now that*

$$\frac{b}{a}(a, -a) + (b, -b) = (0, 0)$$

contradicting the fact that $y^2 - y^1, y^3 - y^1$ *are LI.*

If one or more of the inequalities describing a polyhedron hold at equality, then the polyhedron is not full dimensional. If a polyhedron is not full dimensional, then at least one of the inequalities describing it hold at equality.

More generally, if we take a set $S \subset \mathbb{R}^n$ and a half-space, H, then $S \cap H$ will have a dimension one less than S.

3.4 Properties of convex sets

Theorem 3.23 (Krein–Millman theorem) *Let $S \subset \mathbb{R}^n$ be compact, convex and K the convex hull of its extreme points. Then $S = K$.*

Proof The proof will be by induction on the dimension, n of the space. It is clearly true for $n = 1$. Suppose true for $n - 1$, and $S \setminus K \neq \varnothing$.

Otherwise $K = S$. Since K is closed, by the strict separating hyperplane theorem there is a hyperplane (h, β) that separates y from K, i.e., $h \cdot y < \min_{x \in K} hx$.

Let $c = \min_{x \in S} h \cdot x$. By the Weierstrass theorem c exists and arg $\min_{x \in S} h \cdot x \subset S$. The hyperplane $\{x \colon h \cdot x = c\}$ is disjoint from K, contains S in one of its half-spaces and contains at least one boundary point of S.

Now H is convex and so $H \cap S$ is convex. Further H is closed so, $H \cap S$ is closed and since S is bounded so is $H \cap S$. However $H \cap S$ exists in dimension $n - 1$ and the induction hypothesis applies. So, $H \cap S$ has extreme points and every point in $H \cap S$ is in the convex hull of these extreme points.

It remains to prove that every extreme point of $H \cap S$ is an extreme point of S, suppose not. Let x be an extreme point of $H \cap S$. Since x is not an extreme point of S exists $y, z \in S$ such that $x = \lambda y + (1 - \lambda)z$ for $\lambda \in (0, 1)$. Then

$$c = h \cdot x = \lambda h \cdot y + (1 - \lambda)h \cdot z \geq c.$$

Thus $h \cdot y = h \cdot z = c$, i.e., y, $z \in H \cap S$ contradicting the fact that x is an extreme point of $H \cap S$. ∎

Lemma 3.24 (Intersection lemma) *Let $C^1, C^2, \ldots, C^m \subset \mathbb{R}^n$ be non-empty, compact, convex sets such that $\cup_{j=1}^m C^j$ is convex. If the intersection of any $m - 1$ of them is non-empty then $\cap_{j=1}^m C^j \neq \varnothing$.*

Proof The proof is by induction. Start with the base case of $m = 2$. If $C^1 \cap C^2 \neq \varnothing$ we are done. Otherwise, by Theorem 3.7 there is a hyperplane, H, that strictly separates C^1 from C^2. In particular $H \cap C^1 = \varnothing$ and $H \cap C^2 = \varnothing$. Pick an $x^1 \in C^1$ and $x^2 \in C^2$. Since x^1 and x^2 lie on either side of H the line segment that joins them must pass through H. Since $C^1 \cup C^2$ is convex, this line segment lies in $C^1 \cup C^2$, contradicting the fact that H separates C^1 from C^2.

Now suppose the lemma is true for all $m \leq r$ for some $r > 2$. We show that it must be true for $m = r + 1$. Let $K = \cap_{j=1}^r C^j$. By the induction hypothesis, $K, C^{r+1} \neq \varnothing$. If $K \cap C^{r+1} \neq \varnothing$ we are done. So, for a contradiction, suppose otherwise. Since K and C^{r+1} are compact and convex there is by Theorem 3.7 a hyperplane H that strictly separates them. In particular $C^{r+1} \cap H = \varnothing$.

Set $K^j = C^j \cap H$ for all j. We show that $K^1, K^2, \ldots, K^r$ satisfy the hypothesis of the lemma. First,

$$\bigcup_{j=1}^r K^j = \bigcup_{j=1}^r [C^j \cap H] \cup [C^{r+1} \cap H] = \left[\bigcup_{j=1}^{r+1} C^j \right] \cap H.$$

Since $\cup_{j=1}^{r+1} C^j$ and H is convex, their intersection is convex so $\cup_{j=1}^{r} K^j$ is convex.

Now the intersection of any r of $\{C^1, C^2, \ldots, C^r\}$ overlaps with K and C^{r+1} and therefore with H. Thus any $r - 1$ of $\{K^1, K^2, \ldots, K^r\}$ have non-empty intersection. Therefore, by the induction assumption, $\cap_{j=1}^{r} K^r = K \cap H \neq \emptyset$, which contradicts the fact that H strictly separates K from C^{r+1}. This contradiction proves the result. $\blacksquare$

Theorem 3.25 (Helly's theorem)[4]   *Let $C^1, C^2, \ldots, C^m$ be non-empty, compact convex subsets of $\mathbb{R}^n$ where $m \geq n + 1$. If the intersection of any $n + 1$ of them is non-empty then $\cap_{j=1}^{m} C^j \neq \emptyset$.*

Proof   We prove something a little stronger. If every subset of $\{C^1, C^2, \ldots, C^m\}$ of size $r \geq n + 1$ has non-empty intersection then each collection of size $r + 1$ has non-empty intersection.

Set $K^{-i} = \cap_{j \leq r+1, j \neq i} C^j$ for $i = 1, 2, \ldots, r + 1$. By assumption each K^{-i} is non-empty. For each $i = 1, 2, \ldots, r + 1$ choose a $x^i \in K^{-i}$ and let $S = \text{conv}(x^1, x^2, \ldots, x^{r+1})$. Clearly $S \subseteq \cup_{j=1}^{r+1} C^j$. Set $T^j = C^j \cap S$. Notice that each T^j is, non-empty, compact and convex. Furthermore $\cup_{j=1}^{r+1} T^j = S$, i.e., the union of the T^j's is convex. Therefore the conditions of the intersection lemma are satisfied by the T^j's. Hence $\cap_{j=1}^{r+1} T^j \neq \emptyset$ which implies that $\cap_{j=1}^{r+1} C^j \neq \emptyset$. $\blacksquare$

Any convex set can be continuously transformed into any other convex set of the same dimension.

Definition 3.26   *A set A is **topologically equivalent** to a set B if there exists a continuous function g with continuous inverse such that $g(A) = B$ and $g^{-1}(B) = A$.*

The **closed n-ball** of center c in $\mathbb{R}^n$ is the set $\{x \in \mathbb{R}^n : d(x, c) \leq 1\}$. Note that a closed n-ball is of dimension n.

Theorem 3.27   *A non-empty compact convex set $S \subset \mathbb{R}^n$ of dimension $m \leq n$ is topologically equivalent to a closed ball in $\mathbb{R}^m$.*

Proof   Since S is of dimension m we can find $m + 1$ affinely independent vectors, $\{x^0, x^1, \ldots, x^m\}$ in S. Let $c = (\sum_{j=0}^{m} x^j)/(m + 1)$. By convexity of S, $c \in S$.

Let $K = \text{span}(x^1 - x^0, x^2 - x^0, \ldots, x^m - x^0)$ and H the hyperplane consisting of all points that can be written as $c + z$ where $z \in K$. Observe that $S \subset H$.

Let B be a closed m-ball centered at c. Notice that $B \subset H$. Every point in B can be described in terms of its distance μ and direction u from c. Direction can always be specified in terms of a unit vector in K and distance from c is a number in $[0, 1]$. We show that a similar description is possible for every point in S.

For each unit vector u in K, let $\rho(u)$ be the largest positive number such that $c + \rho(u)u \in S$. In words, $\rho(u)$ is the distance from c along u to the boundary of S. Convexity and compactness of S make this well defined.[5]

We show that $\rho(u)$ is continuous in u. Suppose a convergent sequence $\{u^r\}_{r \geq 1}$ with limit u^*. We must show that $\rho(u^r) \rightarrow \rho(u^*)$. Suppose not. The sequence $\{\rho(u^r)\}$ is bounded, so by the Bolzano-Weierstrass theorem, it contains a convergent subsequence with limit $\rho^* \neq \rho(u^*)$, say.

Consider $\{c + \rho^* u^r\}_{r \geq 1}$ and $\{c + \rho(u^*)u^r\}_{r \geq 1}$. Each is a sequence of boundary points with a limit that must also be a boundary point of S. The limits are $c + \rho^* u^*$ and $c + \rho(u^*)u^*$ respectively. But this implies two boundary points in the same direction, u^*, from c which cannot be.

Given the ρ function, each point $x \in S$ can be expressed as $c + \mu\rho(u)u$ for some unit vector u and $\mu \in [0, 1]$. As in the ball, each point is described by a direction, u, and a distance as measured by the fraction of the total distance to the boundary from c in direction u.

We construct the continuous function g and its inverse as follows. To each point (μ, u) in B we associate a point $c + \mu\rho(u)u$ in S. To each point $x = c + \mu\rho(u)u$ in S we associate the point (μ, u) in B. Continuity follows from continuity of ρ. ∎

3.5 Application: linear production model

We consider a very simple model of an economy in which all relationships between input and output are linear. The ingredients of the model are listed below:

- A non-negative input vector $x \in \mathbb{R}^m$.
- A non-negative output vector $y \in \mathbb{R}^n$.
- An $m \times n$ production matrix, P that relates output to inputs as follows: $y = xP$. Here p_{ij} is the amount of the jth output generated from one unit of the ith input.
- A non-negative resource/capacity vector $b \in \mathbb{R}^k$ that lists the amount of raw materials available for production.
- An $m \times k$ non-negative consumption matrix C that relates inputs to resources: $xC \leq b$. Here c_{ij} is the amount of resource j consumed to produce one unit of input i.

The **input space** is simply $X = \{x \in \mathbb{R}^m : xC \leq b, \ x \geq 0\}$ and the output space is $Y = \{y \in \mathbb{R}^n : y = xP, \ x \in X, \ y \geq 0\}$.

Lemma 3.28 *There is a matrix D with n rows and a vector r such that $Y = \{y \in \mathbb{R}^n : yD \leq r\}$.*

Proof The set X is a polyhedron. In fact it is a polytope since X has no rays. This follows from the fact that C is non-negative and that $xC \leq b$.

Let $x^1, x^2, \ldots, x^k$ be the extreme points of X. Pick any $y \in Y$. Then there is an $x \in X$ such that $y = xP$. Since $x \in X$, x can be written as a convex combination of extreme points of X, i.e.,

$$x = \lambda_1 x^1 + \lambda_2 x^2 + \cdots + \lambda_k x^k.$$

Thus

$$y = \lambda_1 x^1 P + \lambda_2 x^2 P + \cdots + \lambda_k x^k P.$$

In other words, each $y \in Y$ can be written as a convex combination of $\{x^1 P, x^2 P, \ldots, x^k P\}$. It is easy to see that any convex combination of these points is also in Y. Hence Y is a convex combination of a finite number of points. Since Y is a polytope, it is a polyhedron and the lemma follows. ∎

An output vector y^* is called **efficient** if there is no other $y \in Y$ such that $y \geq y^*$.

Theorem 3.29 *A vector $y^* \in Y$ is efficient iff there exists a non-negative, non-trivial price vector p such that $y^* \cdot p \geq y \cdot p$ for all $y \in Y$.*

Proof If $y^* \cdot p \geq y \cdot p$ $\forall y \in Y$ and some price vector p then y^* is clearly efficient. Suppose now that y^* is efficient.

From Lemma 3.28, we know that there is a matrix D and vector r such that $Y = \{y \in \mathbb{R}^n: yD \leq r\}$. Let $S = \{j: y^* \cdot d^j = r_j\}$ where d^j is the jth column of the matrix D. We show that $S \neq \emptyset$. Suppose not. Then $y^* \cdot d^j < r_j$ for all j. Let w be the vector obtained from y^* by adding $\epsilon > 0$ to the first component of y^*. Then $w \cdot d^j = y^* \cdot d^j + \epsilon d_{1j}$. The assumption that $S = \emptyset$ allows us to choose ϵ sufficiently small so that $y^* \cdot d^j + \epsilon d_{1j} \leq r_j$. Thus $w \in Y$ and $w \geq y^*$ violating the efficiency of y^*.

Consider now the system $\{z \cdot d^j \leq 0\}_{j \in S}$. We claim that there is no non-trivial non-negative solution z. If not, there is an $\epsilon > 0$ sufficiently small such that

$$(y^* + \epsilon z)d^j \leq r_j \ \forall j,$$

implying that $(y^* + \epsilon z) \in Y$, contradicting the efficiency of y^*.

Since the system $\{z \cdot d^j \leq 0\}_{j \in S}$ does not admit a non-trivial non-negative solution, we have by the Farkas lemma, non-negative numbers $\{\lambda_j\}_{j \in S}$ such that $\sum_{j \in S} \lambda_j d^j > 0$. Setting $p = \sum_{j \in S} \lambda_j d^j$ completes the proof. ∎

Problems

3.1 Prove the following facts about convex sets:

1. The set $F = \{x: Ax = b, \ x \geq 0\}$ is convex.
2. If C is convex show that $\alpha C = \{y: y = \alpha x, x \in C\}$ is convex for all real α.
3. If C and D are convex sets, then the set $\{y: y = x + z, \ x \in C, \ z \in D\}$ is convex.

4. The intersection of any collection of convex sets is convex.
5. Prove or disprove: the union of convex sets is convex.

3.2 Let C be a convex set and b not in the closure of C. Show that there is vector h such that $hb < \inf_{x \in C} hx$.

3.3 In the plane draw the cone C generated by the *infinite* sequence of vectors $a^j = (j, 1)$, $j = 1, 2, 3, \ldots$

1. Is C closed?
2. Let $b = (1, 0)$. Is $b \in C$? If not, is there a point in C closest to b?
3. Let A be the $2 \times \infty$ matrix whose jth column is a^j. Does the Farkas lemma hold for A?

3.4 If $S \subset \mathbb{R}^n$ is finite, show that conv(S) is a closed set. Is this statement still true if S is not finite?

3.5 Sketch the convex hull of the following set: $\{(x_1, x_2): x_2 = x_1^2, 0 \le x_1 \le 1\}$.

3.6 Let $S \subset \mathbb{R}^n$ be finite and $b \notin$ conv(S). Prove that

$$\max_{x \in \text{conv}(S)} d(x, b) = \max_{x \in S} d(x, b).$$

Show by example that the following is false:

$$\min_{x \in \text{conv}(S)} d(x, b) = \min_{x \in S} d(x, b)$$

3.7 Let A be a $m \times n$ matrix $K = \{y: \text{s.t. } y = Ax,\ \|x\| \le 1\}$. Show that K is convex.

3.8 Let A be a $m \times n$ matrix and $b \in \mathbb{R}^m$. Show that $\{x: Ax = b\} \cap \{x: \|x\| \le 1\} \ne \varnothing$ iff for all non-trivial $u \in \mathbb{R}^m$,

$$u \cdot b \le \max\{u \cdot Ax: \|x\| \le 1\}.$$

3.9 Let $P = \{(x, y): Ax + By \le b\}$ where $x \in \mathbb{R}^n$, $y \in \mathbb{R}^k$, $b \in \mathbb{R}^m$, A is a $m \times n$ matrix and B is a $m \times k$ matrix. Assume $P \ne \varnothing$. Let $Q = \{x \in \mathbb{R}^n: \exists y \in \mathbb{R}^k \text{ s.t. } (x, y) \in P\}$.

1. Suppose that $uB = 0, u \ge 0$ has a non-trivial solution. Use the Farkas lemma to show that Q is defined by the following collection of inequalities:

$$\{uAx \le ub: u \ge 0,\ uB = 0,\ u \ne 0\}$$

2. Suppose that the only solution to $uB = 0$, $u \ge 0$ is the trivial one. Show that $Q = \mathbb{R}^n$.

Notes

1 You should convince yourself that the limit of x^n is non-negative.
2 Hermann Minkowski (1864–1909) was a teacher of Albert Einstein of whom he wrote: 'The mathematical education of the young physicist [Albert Einstein] was not very solid, which I am in a good position to evaluate since he obtained it from me in Zurich some time ago'.

Einstein once referenced some of Minkowski's work in a lecture in this way: 'This has been done elegantly by Minkowski; but chalk is cheaper than grey matter, and we will do it as it comes.'

Hermann Klaus Hugo Weyl (1885–1955) is more famous for his contributions to Quantum Mechanics. Of taxes he once observed, 'Our federal income tax law defines the tax y to be paid in terms of the income x; it does so in a clumsy enough way by pasting several linear functions together, each valid in another interval or bracket of income'.
3 Constantin Caratheodory (1873–1950). Though Greek, he was born in Germany and raised in Brussels. He did spend a brief portion of his life in Greece, where he was instrumental in saving the library at the University of Smyrna when Smyrna was burnt by the Turks in 1922.
4 Eduard Helly (1884–1943). A year into the First World War, 1915, he was shot and captured by the Russians. Though the war ended in 1918, he was not released. It took him another two years to get home to Austria via Japan and Egypt. The bullet destroyed his health while the war did the same for his mathematical career.
5 To see why convexity matters, suppose S were an annulus and c its center.

References

Eggleston, H. G.: 1958, *Convexity*, Cambridge tracts in mathematics and mathematical physics, no. 47, University Press, Cambridge [Eng.].
Lay, S. R.: 1992, *Convex sets and their applications*, Krieger Pub. Co., Malabar, FL.

4 Linear programming

The problem of optimizing a linear function subject to linear inequality and equality constraints is called linear programming (LP). Here is an example of an LP:

$$
\begin{array}{rl}
\max & x_1 + 2x_2 \\
\text{s.t.} & x_1 + \tfrac{8}{3}x_2 \leq 4, \\
& x_1 + x_2 \leq 2, \\
& 2x_1 \leq 3, \\
& x_1, x_2 \geq 0.
\end{array}
$$

Here 's.t.' is an abbreviation for 'subject to'. The set of variables that satisfy the constraints forms a polyhedron. This polyhedron, the shaded part of Figure 4.1, is called the **feasible region** of the LP. In this case the feasible region is a polytope.

A geometrical rendition of our optimization problem is to find a point in the feasible region that maximizes $f(x_1, x_2) = x_1 + 2x_2$. Observe that the optimal solution cannot be in the interior of the feasible region. Suppose it were. Call it (a, b). Let $\epsilon > 0$ be sufficiently small such that $(a + \epsilon, b + \epsilon)$ is feasible. Such an ϵ exists because (a, b) is in the interior of the feasible region. Notice that

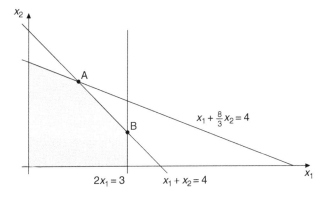

Figure 4.1

$f(a + \epsilon, b + \epsilon) > f(a, b)$, contradicting the optimality of (a, b). Therefore that the optimal solution must lie on the boundary of the feasible region. In fact we can conclude more: one of the extreme points of the feasible region must be an optimal solution. To illustrate, suppose there is an optimal solution in the interior of the boundary between the points A and B marked on the figure. Call it (a, b). Since this point is on the boundary our previous argument does not apply because $(a+\epsilon, b+\epsilon)$ need not be feasible. The idea is to perturb (a, b) to a new feasible point that is still on the same boundary segment. Consider the point $(a + \mu_1, b + \mu_2)$. We want this to be on the same boundary segment that (a, b) is on. That boundary is defined by the equation $x_1 + x_2 = 2$. So we need $a + \mu_1 + b + \mu_2 = 2$. Since $a + b = 2$ it follows that $\mu_1 + \mu_2 = 0$. We must ensure that the μ_1 and μ_2 are chosen so that $(a + \mu_1, b + \mu_2)$ is feasible. Given the location of (a, b) we know that all the other inequalities are satisfied strictly. That is $a + 8/3b < 4$, $2a < 3$ and $a, b > 0$. So, for $|\mu_1|, |\mu_2|$ sufficiently small $(a+\mu_1, b+\mu_2)$ will be feasible.

Notice that $f(a + \mu_1, b + \mu_2) = a + 2b + \mu_1 + 2\mu_2 = a + b + \mu_2$ because $\mu_1 = -\mu_2$. If we choose $\mu_2 > 0$ then $f(a+\mu_1, b+\mu_2) > f(a, b)$ which contradicts the optimality of (a, b).

In this case, the optimal solution is at the point A (which the reader should verify). It is formed by the intersection of the lines $x_1 + x_2 = 2$ and $x_1 + 8/3x_2 = 4$.

If an LP has inequality constraints, the constraints that are satisfied at equality by a feasible solution are said to **bind** at that solution. In our example, the constraints $x_1 + x_2 \leq 2$ and $x_1 + 8/3x_2 \leq 4$ bind at an optimal solution. They will be called (when there is no ambiguity) **binding constraints**. The function cx being optimized is called the **objective function** and the matrix A defining the feasible region is called the **constraint matrix**. The vector b is called the **vector of right-hand sides** or **RHS** for short.

Every linear programming problem can be written in the following standard form:

$$\begin{aligned} \max \ & cx \\ \text{s.t.} \ & Ax = b \\ & x \geq 0 \end{aligned}$$

To convert any LP into into this form the following modifications listed below are performed:

- If x is unrestricted then substitute $x_j = x_j^+ - x_j^-$ $x_j^+, x_j^- \geq 0$.
- If a constraint is in the form $\sum_{j=1}^{n} a_{ij}x_j \leq b_i$ then add a **slack** variable $s_i \geq 0$ such that $\sum_{j=1}^{n} a_{ij}x_j + s_i = b_i$.
- If a constraint is in the form $\sum_{j=1}^{n} a_{ij}x_j \geq b_i$ then subtract a **surplus** variable $s_i \leq 0$ such that $\sum_{j=1}^{n} a_{ij}x_j - s_i = b_i$.
- If the objective is min cx then replace it with it: max $-cx$.

- To change $\sum_{j=1}^{n} a_{ij}x_j = b_i$ to an inequality constraint, replace equality with these two sets of inequality constraints: $\sum_{j=1}^{n} a_{ij}x_j \leq b_i$ and $-\sum_{j=1}^{n} a_{ij}x_j \leq -b_i$.

The standard form of the LP above is

$$
\begin{array}{llll}
\max & x_1 + 2x_2 & & \\
\text{s.t.} & x_1 + \frac{8}{3}x_2 + s_1 & = 4, \\
& x_1 + x_2 + s_2 & = 2, \\
& 2x_1 + & s_3 = 3, \\
& x_1, x_2, s_1, s_2, s_3 \geq 0.
\end{array}
$$

Exactly one of three things will be true of every LP:

1. It is **infeasible**, meaning that there is no solution to $\{x \in \mathbb{R}_+^n : Ax = b\}$. As an example consider $\max\{x : \text{s.t. } x \leq 5, x \geq 6\}$.
2. The optimal objective function value is **unbounded**. That is for all positive real numbers t there is a $z \in \{x \in \mathbb{R}_+^n : Ax = b\}$ such that $c \cdot z \geq t$. As an example consider $\max\{x : \text{s.t. } x \geq 6\}$. It is important to distinguish between an LP that is unbounded and one that has an unbounded feasible region. An LP with unbounded objective function value will have a unbounded feasible region. The converse is not true. The following LP: $\min\{x : x \geq 3\}$ has an unbounded feasible region but does not have an unbounded optimal objective function value.
3. It has a **finite** optimal objective function value. As an example consider $\max\{x : \text{s.t. } x \leq 5\}$. Note that an LP can have multiple optimal solutions. For example $\max\{x_1 + x_2 : \text{s.t. } x_1 + x_2 \leq 1\}$.

To get a sense of the importance of the subject of this chapter, we recount the following from Nicholas Hall[1] while teaching a class on linear programming. 'A student of mine once prefaced his request for a grade change with the observation that three important things had come out of Second World War. The first was women in the workforce, the second was the atomic bomb and the third was linear programming.'

The reader can contact Professor Hall to discover the fate of the students request.

The subject of linear programming is older than the Second World War. Joseph Fourier (1768–1830), of 'series' fame, was amongst the first to investigate this subject and point outs its importance to mechanics and probability theory. The problem that attracted his attention was that of finding a least maximum deviation fit to a system of linear equations. He reduced the problem to that of finding the lowest point of a polyhedron.[2] His suggested solution to this problem can be viewed as a precursor to the modern day simplex algorithm devised by George Dantzig in 1947. Dantzig at the time was engaged in project SCOOP (Scientific Computation of Optimum Programs), an American research program that resulted from the intensive scientific activity during the

Second World War, aimed at rationalizing the logistics of the war effort. In the Soviet Union, Leonid Kantorovitch (1912–1986) had already proposed a similar method for the analysis of economic plans, but his contribution remained unknown to the general scientific community until much later. Kantorovitch but not Dantzig was awarded the Nobel prize for the development of linear programming.[3]

4.1   Basic solutions

Consider the standard form LP, $\max\{cx\colon Ax = b, x \geq 0\}$. Assume it to be feasible. Before proceeding we describe a standard argument that allows us to suppose that the constraint matrix A has rank m and that $n \geq m$. Suppose that $b \in \mathrm{span}(A)$, otherwise the LP is infeasible and our story ends. Consider the augmented matrix $[A|b]$. Since $b \in \mathrm{span}(A)$, the rank of A and $[A|b]$ coincide.

If the rank of $[A|b]$ is less than m, it means that some row of $[A|b]$ is a linear combination of other rows of $[A|b]$. In other words one of the equations in the system $Ax = b$ is implied by a linear combination of the others. This equation is redundant and can be eliminated without changing the set of solutions.

If $m > n + 1$, then $[A|b]$ must have at least one redundant row. This is because the rows of $[A|b]$ as vectors live in a space of dimension $n + 1$. Thus any set of at least $n + 2$ vectors must be LD. There must be a redundant equation and we can delete the corresponding row of $[A|b]$. This process can be repeated as long as the number of rows of $[A|b]$ exceeds the number of its columns. Therefore, we may suppose that both the number of rows as well as the rank of $[A|b]$ coincide and cannot exceed $n + 1$. Thus the rank of A can be assumed to be m. Again, thinking of the rows as vectors in $\mathbb{R}^n$, since we have m LI row vectors, $m \leq n$.

In this section we derive an algebraic characterization of the extreme points of $\{x\colon Ax = b, x \geq 0\}$.

Definition 4.1   *Given $Ax = b$, where A is an $m \times n$ matrix, let B be a $m \times m$ non-singular submatrix of A. B is called a **basis** of A. Let the rest of the matrix A be submatrix N; then $A_{m*n} = [B_{m*m} | N_{m*(n-m)}]$.*

Variables associated with the columns of B will be called **basic**, and the others **non-basic**.

Definition 4.2   *Let B be a basis for A. Set $x_j = 0$ if $j \in N$. For x_j s.t. $j \in B$, choose them so as to solve $Bx^B = b$. Notice the choice will be unique because B is a non-singular square matrix. The resulting solution is called a **basic solution**.*

Definition 4.3   *If a basic solution x associated with the basis B, $x = [x^B | 0] = [B^{-1}b | 0]$, is non-negative then x is a **basic feasible solution** to the LP.*

Example 15 *Consider the system*

$$
\begin{aligned}
x_1 + x_2 + x_3 &= 1, \\
2x_1 + 3x_2 \quad\ &= 1,
\end{aligned}
$$

$$x_1, x_2, x_3 \geq 0.$$

The constraint matrix is

$$
\begin{bmatrix}
1 & 1 & 1 \\
2 & 3 & 0
\end{bmatrix}.
$$

Here is one basis:

$$
\begin{bmatrix}
1 & 1 \\
2 & 0
\end{bmatrix}.
$$

To find the basic solution associated with this basis, we set $x_2 = 0$ and solve

$$
\begin{aligned}
x_1 + x_3 &= 1, \\
2x_1 + 0x_3 &= 1.
\end{aligned}
$$

So, the basic solution is $x_1 = 1/2$, $x_2 = 0$ and $x_3 = 1/2$, which also happens to be a basic feasible solution.
Yet another basis is

$$
\begin{bmatrix}
1 & 1 \\
2 & 3
\end{bmatrix}.
$$

The basic solution associated with this basis is found by setting $x_3 = 0$ and solving

$$
\begin{aligned}
x_1 + x_2 &= 1, \\
2x_1 + 3x_2 &= 1.
\end{aligned}
$$

The basic solution is $x_1 = 2$, $x_2 = -1$ and $x_3 = 0$ which is not a basic feasible solution.

Lemma 4.4 *If the set $\{x\colon Ax = b, x \geq 0\}$ is feasible, then it has a basic feasible solution.*

Proof Since the LP is feasible, $b \in \mathrm{cone}(A)$. From the proof of Lemma 3.8, we know that b can be expressed as a non-negative linear combination of LI columns of A, B say. If these columns form a basis we are done. If not, since A is of full rank, we can augment B with additional columns to form a basis. The x variables associated with these columns would be set to zero, this completes the proof.

For the reader who skipped the proof of Lemma 3.8 a proof is provided below. Let x' be a feasible solution. Then $b_i = \sum_{j \in S} a_{ij} x'_j$ where $S = \{j: x'_j > 0\}$. We ignore terms in $\{j: x'_j = 0\}$ since they are zero. If $\{a^j: j \in S\}$ are linearly independent we are done.[4] If the cardinality of this set is less than m, throw in some additional columns of the A matrix to produce a set of m LI vectors. We can do this because of the full rank assumption. The variables associated with these extra columns take the value zero. Then x' is a basic feasible solution.

Assume $\{a^j: j \in S\}$ are not linearly independent. Then there exists $\{\lambda_j\}$ not all zero s.t. $\sum_{j \in S} \lambda_j a^j = 0$. Let $x^{\text{new}} = x' - \theta \lambda \geq 0$ by picking θ as small as necessary. The columns of A associated with the positive components of x^{new} involve one fewer dependent column. Next, we verify that x^{new} is feasible.

$$Ax^{\text{new}} = A(x' - \theta \lambda) = Ax' - \theta A\lambda$$

$$= Ax' - \theta \sum_{j \in S} \lambda_j * a^j = Ax' - \theta * 0 = Ax' = b.$$

If the columns associated with the non-zero components of x^{new} are LD, repeat the argument above. As there are finite number of columns and the method eliminates one column at each iteration, it will terminate after a finite number of steps. ∎

Lemma 4.5   *If x^* is a basic feasible solution of the set $\{x: Ax = b, x \geq 0\}$, then x^* is an extreme point of the set.*

Proof   If x^* is not an extreme point there exist y and z feasible, distinct from x^*, such that $x^* = \lambda y + (1 - \lambda)z$. Let B be the basis associated with x^* and set $x^* = [x^B | x^N]$, $A = [B|N]$, $y = [y^B | y^N]$, $z = [z^B | z^N]$.

From definitions $\lambda y^N + (1 - \lambda)z^N = x^N = 0 \Rightarrow y^N = z^N = 0 = x^N$. Feasibility implies

$$Ay = b \Rightarrow By^B = b$$

and

$$Az = b \Rightarrow Bz^B = b.$$

But x^B is the unique solution to $Bx = b$, then $x^B = z^B = y^B$, so $x^* = z = y$. As a result there do not exist z, y different than x^*. Therefore x is an extreme point. ∎

The non-negativity restriction is crucial. The system $x_1 + x_2 = 4$ has basic solutions, but no extreme points.

Lemma 4.6   *Every extreme point of the set $\{x: Ax = b, x \geq 0\}$ is a basic feasible solution.*

Proof Let x^* be an extreme point and let $B = \{a^j : x_j^* > 0\}$ and $N = \{a^j : x_j^* = 0\}$. If B is invertible, then we are done. If B is not invertible, there exists $y^B \neq 0$ such that $By^B = 0$. Let $y = [y^B | y^N]$ where $y^N = 0$.

Define $x^1 = x^* + \theta y \geq 0$ and $x^2 = x^* - \theta y \geq 0$ by choosing θ small enough. x^1 and x^2 are feasible because $Ay = 0$ and $Ax^* = b$. But this contradicts the fact that x^* is an extreme point since

$$x^* = \tfrac{1}{2}x^1 + \tfrac{1}{2}x^2.$$

∎

Theorem 4.7 (Fundamental theorem of linear programming) *Let $P = \{x : Ax = b, x \geq 0\}$. If A is of full row rank and $\max_{x \in P} cx$ has a finite optimal solution, there is an optimal solution at one of the extreme points of P.*

Proof From the Resolution theorem we can express P as $Q + C$ where Q is a polytope and C a cone. By Lemma 4.4 we know that P has at least one extreme point. Therefore Q will be the convex hull of the extreme points of P. Every $x \in P$ can be expressed as a convex combination of extreme points of Q and a non-negative linear combination of the extreme rays of C. Let $\{e^t\}_{t \geq 1}$ be the set of extreme points of Q and $\{r^k\}_{k \geq 1}$ the extreme rays of C. Let x^* be an optimal solution of the LP. Then

$$x^* = \sum_{t \geq 1} \lambda_t e^t + \sum_{k \geq 1} \mu_k r^k,$$

where $\lambda_t \geq 0$ for all $t \geq 1$, $\mu_k \geq 0$ for all $k \geq 1$ and $\sum_{t \geq 1} = 1$.

We prove that we can choose an optimal solution x^* such that $\mu^k = 0$ for all k. Without loss of generality, suppose that $\mu_1 > 0$. We have three cases.

Case 1: If $c \cdot r^1 > 0$, then the solution

$$x' = \sum_t \lambda_t e^t + (\mu_1 + \delta)r^1 + \sum_{k \geq 2} \mu_k r^k$$

where $\delta > 0$ has an objective function value $c \cdot x' > c \cdot x^*$ which contradicts the optimality of x^*.

Case 2: If $c \cdot r^1 < 0$ repeat the argument above with $\delta < 0$.

Case 3: If $c \cdot r^1 = 0$, consider the vector $x' = \sum_{t \geq 1} \lambda_t e^t + \sum_{k \geq 2} \mu_k r^k$. It is clearly feasible and given the hypothesis of this case, $c \cdot x^* = c \cdot x'$, it is optimal. Now repeat the argument with x^* replaced by x'. Hence

$$x^* = \sum_{t \geq 1} \lambda_t e^t.$$

Therefore,

$$c \cdot x^* = \sum_{t \geq 1} \lambda_t (c \cdot e^t).$$

Thus $c \cdot x^*$ is a weighted average of numbers of of the form $\{c \cdot e^t\}_{t \geq 1}$. However each $c \cdot e^t \leq cx^*$ which means $c \cdot e^t = c \cdot x^*$ for at least one t. Since e^t is an extreme point of P, the proof is complete. ∎

4.2 Duality

Associated with each LP is another LP called its **dual**. The original LP is called the **primal**.[5] To motivate the dual consider the following non-negative combination of inequalities in the example from the beginning of this chapter:

$$\frac{3}{5}(x_1 + \frac{8}{3}x_2 \leq 4)$$
$$+ \quad \frac{2}{5}(x_1 + \quad x_2 \leq 2)$$
$$\overline{\qquad x_1 + 2x_2 \leq \frac{16}{5}}$$

As a result $16/5$ is an upper bound on the objective function value of the example problem. Such upper bounds on the optimal objective function value can be found by taking appropriate linear combinations of constraints (yA) that dominate the objective function c, i.e., $c \leq yA \Rightarrow cx \leq yAx$ since $x \geq 0$. Using the fact that $Ax = b$ allows one to conclude that

$$cx \leq yAx = yb \quad \Rightarrow \quad cx \leq yb.$$

Thus yb is an upper bound on the objective function value. The problem of finding the smallest such upper bound is called the dual.

Primal (P)		Dual (D)
$Z_P = \max \ cx$		$Z_D = \min \ yb$
s.t. $\quad Ax = b$	$\Longrightarrow$	s.t. $\quad yA \geq c$
$x \geq 0$		y unrestricted.

It follows from the way the dual was motivated that $Z_D \geq Z_P$. This is known as **weak duality**. As an example, we derive the dual to the example problem above. First, we introduce slack variables to produce an equality constrained version of the problem.

$$
\begin{array}{llllllll}
\max & x_1 & + & 2x_2 & + & & + & & + & & \\
\text{s.t.} & x_1 & + & \tfrac{8}{3}x_2 & + & s_1 & + & & + & & = 4 \\
& x_1 & + & x_2 & + & & + & s_2 & + & & = 2 \\
& 2x_1 & + & & & & + & & + & s_3 & = 3 \\
\end{array}
$$
$$
x_1, x_2, s_1, s_2, s_3 \geq 0
$$

The dual of the example problem will be

$$
\begin{array}{lllllll}
\min & 4y_1 & + & 2y_2 & + & 3y_3 & \\
\text{s.t.} & y_1 & + & y_2 & + & 2y_3 & \geq 1, \\
& \tfrac{8}{3}y_1 & + & y_2 & + & & \geq 2, \\
& y_1 & + & & + & & \geq 0, \\
& & + & y_2 & + & & \geq 0, \\
& & + & & + & y_3 & \geq 0. \\
\end{array}
$$

Remarkably, under the right conditions, the smallest upper bound on the optimal objective function value of the primal coincides with the optimal objective function value. We prove this next. First a preliminary lemma.

Lemma 4.8 *If problem (P) is infeasible then (D) is either infeasible or unbounded. If (D) is unbounded then (P) is infeasible.*

Proof   Suppose for a contradiction that (D) has a finite optimal solution, y^*, say. Infeasibility of (P) implies by the Farkas lemma a vector $\hat{y}$ such that $\hat{y}A \geq 0$ and $\hat{y} \cdot b < 0$. Let $t > 0$. The vector $y^* + t\hat{y}$ is a feasible solution for (D) since $(y^* + t\hat{y})A \geq y^*A \geq c$. Its objective function value is $(y^* + t\hat{y}) \cdot b < y^* \cdot b$, contradicting the optimality of y^*. Since (D) cannot have a finite optimal, it must be infeasible or unbounded.

Now suppose (D) is unbounded. By the resolution theorem we can write any solution of (D) as $y + r$ where y is a feasible solution to the dual and r is a ray, i.e., $yA \geq c$ and $rA \geq 0$. Furthermore $r \cdot b < 0$ since (D) is unbounded. By the Farkas lemma, the existence of r implies the primal is infeasible. ∎

Theorem 4.9 (Duality theorem)   *If a finite optimal solution for either the primal or dual exists, then $Z_P = Z_D$. Note: We give two proofs.*

Proof   (First) By the previous lemma if one of Z_P and Z_D is finite so is the other. Let x^* be an optimal solution to the primal. If x is any other feasible solution to the primal it is easy to see that $z = x - x^*$ satisfies $Az = 0$, $c \cdot z \leq 0$ and $x^* + z \geq 0$.

Thus, if x^* is optimal there is no z that satisfies

$$Az = 0,$$
$$-Iz \le x^*,$$
$$-c \cdot z < 0.$$

By the Farkas lemma, the following alternative system is feasible:

$$yA - tI - c = 0,$$
$$t \cdot x^* < 0,$$
$$t \ge 0.$$

Let (y^*, t^*) be a feasible solution to the alternative. Since $t^* \ge 0$ it follows that $y^*A = t^*I + c \ge c$, i.e., y^* is a feasible dual solution. Finally,

$$(y^*A - t^*I - c)x^* = 0 \Rightarrow y^*Ax^* - t^* \cdot x^* - c \cdot x^* = 0 \Rightarrow y^*b < c \cdot x^*.$$

Therefore $Z_D \le Z_P$. However,

$$Z_D = y^* \cdot b = y^*Ax^* \ge c \cdot x^* = Z_P.$$

Hence $Z_P = Z_D$. ∎

Proof (Second) By the previous lemma if one of Z_P and Z_D is finite so is the other. Let x^* be an optimal solution to the primal and y^* an optimal solution to the dual. By weak duality

$$Z_D = y^* \cdot b = y^*Ax^* \ge c \cdot x^* = Z_P.$$

To complete the proof we show that $Z_D \le Z_P$. Pick an $\epsilon > 0$ and consider the system

$$-cx \le -Z_P - \epsilon,$$
$$Ax = b,$$
$$x \ge 0.$$

By the definition of Z_P this is infeasible. So, by the Farkas lemma there is a solution to the following system:

$$-\lambda c + yA \ge 0,$$
$$\lambda(-Z_P - \epsilon) + yb < 0,$$
$$\lambda \ge 0.$$

Let that solution be (λ^*, y^*). We show that $\lambda^* > 0$. Suppose not. Since $\lambda^* \geq 0$ it follows that $\lambda^* = 0$. This implies that

$$y^* A \geq 0,$$
$$y^* b < 0.$$

By the Farkas lemma this implies that the system $Ax = b$ with $x \geq 0$ is infeasible which violates the initial assumption.

Let $y' = y^*/\lambda^*$. Since $\lambda^* > 0$ this is well defined. Also

$$y'A = \frac{y^* A}{\lambda^*} \geq c$$

making y' a feasible solution for the dual problem. Further $y'b < Z_P + \epsilon$. Since y' is feasible in the dual, it follows that

$$Z_P \leq Z_D \leq y'b < Z_P + \epsilon.$$

Since $\epsilon > 0$ is arbitrary it follows that $Z_P = Z_D$. ■

The theorem fails if at least one of the pair of primal and dual programs is infeasible. Consider $\max\{x: \text{s.t. } x = 5, x = 4, x \geq 0\}$. This is clearly infeasible. Its dual is $\min\{5y_1 + 4y_2: \text{s.t. } y_1 + y_2 \geq 0\}$. The dual is feasible but it is also unbounded.

Suppose that in an optimal solution to the dual, y^*, one of the constraints was not binding, i.e., $\sum_{i=1}^{m} a_{ij} y_i^* > c_j$ for some j. Then eliminating this constraint from the dual will not affect the optimal objective function of the dual. Eliminating this constraint would correspond, in the primal, to setting $x_j = 0$. This connection is formalized below.

Theorem 4.10 (Complementary slackness) *If the feasible pair (x^*, y^*) is optimal for the primal and the dual programs, then*

1. $x_j^* > 0 \Rightarrow \sum_{i=1}^{m} a_{ij} y_i^* = c_j$,

2. $\sum_{i=1}^{m} a_{ij} y_i^* > c_j \Rightarrow x_j^* = 0$.

Proof Let (x^*, y^*) be an optimal pair for the primal and dual programs. We will prove the following equivalent statement:

$$\left[\sum_{i=1}^{m} a_{ij} y_i^* - c_j \right] x_j^* = 0 \ \forall j.$$

Stated in vector–matrix notation: $(y^* A - c)x^* = 0$.
From the duality theorem, $y^* b - cx^* = 0$. However $b = Ax^*$. So, $y^* Ax^* - cx^* = 0$, which is the required result. ■

4.3   Writing down the dual

The following table provides rules for constructing a dual problem from a primal problem. If a primal problem has variables $x_1, x_2, \ldots, x_n$, objective function $c \cdot x$ and constraint matrix A, the dual will have variables $y_1, y_2, \ldots, y_m$, one for each constraint, objective function $y \cdot b$ and constraint matrix A^T.

Primal	Dual
$\max c \cdot x$	$\min y \cdot b$
$\sum_j a_{ij} x_j \leq b_i$	$y_i \geq 0$
$\sum_j a_{ij} x_j = b_i$	y_i unrestricted
$\sum_j a_{ij} x_j \geq b_i$	$y_i \leq 0$
$x_j \geq 0$	$\sum_i a_{ij} y_i \geq c_j$
x_j unrestricted	$\sum_i a_{ij} y_i = c_j$
$x_j \leq 0$	$\sum_i a_{ij} y_i \leq c_j$

4.4   Interpreting the dual

The Morpheus[6] company makes two kinds of liquid soporifics: white soma and red soma.[7] Each gallon of white soma can be sold for $1, while each gallon of red soma can be sold for $2. The production capacity of the company limits them to producing a total of 2,000 gallons of soma. Each gallon of white soma requires 1 hour of labor to process and package. Each gallon of red soma requires 8/3 hours of labor to process and package. The company has a total of 4,000 hours of labor available. Government regulation rations the production of white soma. Morpheus has a license that permits it to produce upto 1,500 gallons of white soma. What mix of white and red soma should be produced to maximize the revenue of the Morpheus company?

The problem of the Morpheus company can be formulated as a linear program. Let x_1 denote the number of gallons of white soma (measured in units of a thousand) to be produced and x_2 be the number of gallons of red soma produced (measured in units of a thousand). Since a non-negative amount must be produced, $x_1, x_2 \geq 0$. Revenue will be $x_1 + 2x_2$. The total amount produced $x_1 + x_2$ must be at most 2, i.e., $x_1 + x_2 \leq 2$. Similarly, the limit on labor time means that $x_1 + (8/3)x_2 \leq 4$. The government constraint requires that $x_1 \leq 1.5$. Summarizing, the problem of the Morpheus company is

$$
\begin{aligned}
\max \quad & x_1 + 2x_2 \\
\text{s.t.} \quad & x_1 + \tfrac{8}{3}x_2 \leq 4, \\
& x_1 + x_2 \leq 2, \\
& x_1 \leq \tfrac{3}{2}, \\
& x_1, x_2 \geq 0.
\end{aligned}
$$

The optimal solution to Morpheus' problem is $x_1 = 4/5$, $x_2 = 6/5$ with a revenue of $16/5$. At this solution both the production capacity and labour hour constraint are binding.

Now suppose that the Narziss[8] company wishes to buy the Morpheus company. It will do so by offering a per unit (a thousand gallons) price for each resource that Morpheus possesses. So, Narziss must decide on a price per unit, $y_1 \geq 0$, for the labor hours, a price per unit $y_2 \geq 0$ for production capacity and $y_3 \geq 0$ the price per unit for the right to produce white soma. Narziss, once it acquires the resources of Morpheus, will be able to produce white and red soma and sell it for the same prices as Morpheus does.

If Morpheus sells the ability to produce a single unit of white soma it must give up one unit of capacity, one unit of labor and one unit of its government approved quota. It will receive in return $y_1 + y_2 + y_3$. For this to be a profitable transaction for Morpheus, $y_1 + y_2 + y_3 \geq 1$. Similarly, for red soma $8/3 y_1 + y_2 \geq 2$. Narziss seeks y_1, y_2, y_3 so as to minimize $4y_1 + 2y_2 + 1.5y_3$, its total purchase price. Therefore Narziss must solve

$$
\begin{aligned}
\min \quad & 4y_1 + 2y_2 + \tfrac{3}{2}y_3 \\
\text{s.t.} \quad & y_1 + y_2 + y_3 \geq 1, \\
& \tfrac{8}{3}y_1 + y_2 + 0y_3 \geq 2, \\
& y_1 + 0y_2 + 0y_3 \geq 0, \\
& 0y_1 + y_2 + 0y_3 \geq 0, \\
& 0y_1 + 0y_2 + y_3 \geq 0.
\end{aligned}
$$

Notice that the problem that Narziss solves is the dual to the problem that Morpheus must solve. A consequence of the duality theorem is that the minimum total price Narziss must pay to give a non-negative profit to Morpheus is exactly the maximum revenue that Morpehus can obtain from the production of soma. This should not come as a surprise. For Morpheus to part with the capability to produce soma, it must receive at least as much money as it makes by producing and selling soma. Narziss on the other hand, should pay no more than the revenue it could generate by acquiring the ability to produce soma.

Our story of Narziss and Morpheus allows us to interpret the dual variables as 'prices' for each of the resources. The interpretation is more than cosmetic. To illustrate consider the following question: would additional production capacity be valuable and if so just how much? At first glance the answer seems yes. As long as each additional amount of soma (of either kind) produced can be sold revenue should increase. However, the additional capacity could be worthless if we don't have the labor hours necessary to produce the additional soma. Notice that red soma is more labor intensive than white soma. By cutting back on red soma production we can free up time to expand production of white soma and so make use of the additional capacity. However, red soma generates more revenue per unit than white soma, so a trade-off calculation must be made to determine the additional revenue, if any, from an expansion of production capacity. Remarkably, the optimal solution to the dual to the Morpheus problem will provide us with the answer.

The optimal solution to the dual is $y_1 = 3/5$, $y_2 = 2/5$ and $y_3 = 0$. Each dual variable represents the per unit increase in objective function value from a 'small' increase in the RHS of the corresponding primal constraint other things held fixed.

Consider the variable y_3. This says that an increase in the government imposed limit on white soma production will not increase revenue. To see that this is sensible, observe that in the current optimal solution this constraint is not binding. Since it is non-optimal to produce to this limit, raising it will not increase production of white soma.

What about a decrease in the limit on white soma production? Up to a point this will not make a difference. The current optimal solution produces 4/5 units of white soma. As long as the government limit exceeds 4/5, the current solution is revenue maximizing. So, for slight changes in the value of the relevant RHS, the value of y_3 gives us the change in optimal objective function value.

To see why the dual variable cannot give us the change in optimal objective function value for any size change, suppose the government limit is set at 1/2. The current optimal solution is no longer feasible. Figure 4.2 shows the new feasible region and the original optimal solution (point A) is no longer within it. The new optimal solution is at point B, $x_1 = 1/2$, $x_2 = 9/8$. Notice there is now a change in optimal objective function value. Why does the value of y_3 at the old optimal solution no longer provide a correct forecast of the change in optimal objective function value? The change in RHS has resulted in a change in the constraints that bind at the optimal solution. The first dual solution is no longer optimal after the change in RHS. To summarize, at optimality, each dual variable represents the per unit change in optimal objective function value for a change in the relevant RHS within some prescribed range other things held fixed. One can use the optimal solution to primal and dual to compute this prescribed range. The upper and lower limits of this range are called the **allowable increase** and **allowable decrease**. It is possible for the range to be zero.

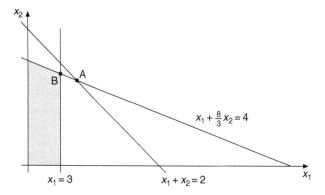

Figure 4.2

Now turn to y_2. It has a value of $2/5$. This means that if we increase production capacity by Δ, for sufficiently small Δ, the optimal objective function value will increase by $(2/5)\Delta$. Let us verify this by increasing the RHS of the second primal constraint (the one associated with productive capacity) by one unit. The new primal optimal solution is $x_1 = 12/5$, $x_2 = 3/5$ with a revenue of $18/5$. This is an increase in revenue of $18/5 - 16/5 = 2/5$.

Since the dual variable represents the rate at which the optimal objective function changes as the relevant RHS changes, it is natural to think of the dual variable as a slope or derivative. To follow this analogy through, consider the following problem:

$$F(b) = \max \quad x_1 + 2x_2$$
$$\text{s.t.} \quad x_1 + \tfrac{8}{3}x_2 \le 4$$
$$x_1 + \quad x_2 \le b$$
$$x_1 \qquad \le \tfrac{3}{2}$$
$$x_1, x_2 \ge 0$$

We raise b from zero and compute $F(b)$. For $b \in [0, 3/2]$, $F(b) = 2b$. For $b \in [3/2, 39/16]$ $F(b) = (2/5)b + 12/5$. For $b \ge 39/16$, $F(b) = 27/8$. Figure 4.3 shows a graph of $F(b)$. This shows that F is piecewise linear in b and non-decreasing. The slope or derivative of F between consecutive break-points is precisely the value of the relevant dual variable at optimality. Break points correspond to changes in the set of constraints binding at optimality. Consider for example $b = 3/2$. For $b < 3/2$, the binding constraints are $x_1 \ge 0$ and $x_1 + x_2 \le b$. For $b > 3/2$ the binding constraints are $x_1 + (8/3)x_2 \le 4$ and $x_1 + x_2 \le 2$. At $b = 3/2$, the dual has two optimal solutions. One where $y_2 = 2$ and the other is $y_2 = 2/5$. The first gives the slope of F for $b < 3/2$ and the other for $b > 3/2$. When $b = 3/2$, there is a choice of values for y_2. The larger, 2,

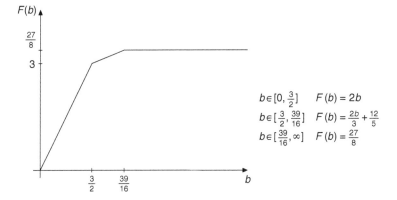

Figure 4.3

gives the reduction in optimal objective function value for a unit reduction in the relevant RHS. The smaller, 2/5, gives the value of an increase in optimal objective function value for a unit increase in the relevant RHS.

The next section formalizes and generalizes the lessons of the example.

4.5 Marginal value theorem

If each constraint of a linear program is interpreted as the limitation imposed by the quantity available of some resource, then the dual variable of that constraint also has an interpretation. The marginal value of that resource is equal to the change in the optimal objective function value from an infinitesimal change in the amount of the resource other things held fixed. The dual variable associated with that constraint is the marginal value of that resource.

Consider the linear program: $\max\{cx: Ax \leq b, x \geq 0\}$ which we will call (P) and its dual (D) $\min\{yb: yA \geq c, y \geq 0\}$. Assume both have feasible, optimal solutions. Fix a $d \in \mathbb{R}^m$ and let

$$f(\epsilon) = \max\{cx: Ax \leq b + \epsilon d, x \geq 0\}$$

for all $\epsilon \geq 0$. If the program is infeasible for some value of ϵ, set $f(\epsilon) = -\infty$. Observe that

$$f(0) = \max\{cx: Ax \leq b, x \geq 0\} = \min\{yb: yA \geq c, y \geq 0\}.$$

Amongst all optimal solutions to (D), call the one that minimizes dy, y^*. Therefore y^* is the solution to

$$\min\{dy: yA \geq c, yb = f(0), y \geq 0\}.$$

Theorem 4.11 $f(\epsilon) \leq f(0) + \epsilon dy^*$ *for all $\epsilon \geq 0$ with equality for all sufficiently small ϵ.*

Proof If $f(\epsilon) = -\infty$ we are done. So we may suppose that the program that defines $f(\epsilon)$ is feasible. By the duality theorem

$$f(\epsilon) = \min\{y(b + d\epsilon): yA \geq c, y \geq 0\}.$$

Since y^* is a feasible solution to this last program it follows that

$$f(\epsilon) \leq y^*(b + d\epsilon) = f(0) + \epsilon dy^*.$$

To complete the proof we show that for all $\epsilon \geq 0$ sufficiently small that

$$f(\epsilon) \geq f(0) + \epsilon dy^*.$$

The dual to the program that defines y^* is

$$\max\{cx - f(0)t: Ax - tb \leq d, (x,t) \geq 0\}.$$

Let (x^*, t^*) be the optimal solution this program. By the duality theorem

$$cx^* - f(0)t^* = dy^*.$$

Let x^0 be an optimal solution to (P). Choose $\epsilon \leq 1/t^*$, when $t^* = 0$, take ϵ to be any positive number. Consider

$$x = (1 - t\epsilon)x^0 + \epsilon x^*.$$

Since $x \geq 0$ and $Ax \leq b + \epsilon d$ it follows that x is a feasible solution to the program that defines $f(\epsilon)$. Hence,

$$
\begin{aligned}
f(\epsilon) \geq cx &= (1 - t\epsilon)cx^0 + \epsilon cx^* \\
&= (1 - t\epsilon)f(0) + \epsilon(dy^* - f(0)t^*) = f(0) + \epsilon dy^*.
\end{aligned}
$$ ∎

4.6 Application: zero-sum games

Definition 4.12 *A* **zero-sum game** *is given by an $m \times n$ matrix A. The two players are called Row and Column. The $\{ij\}$th entry of A, a_{ij}, is the payoff to Row from Column when Row chooses row i and Column chooses column j.*

Rows and columns correspond to what are called **pure strategies**. Strategy choices in the game are simultaneous and the matrix A is known to both players. Players are assumed to care only about expected payoffs (i.e. are risk neutral).

A zero-sum game familiar to many is 'rock, paper, scissors'.[9] Each player has three pure strategies called rock, paper and scissors respectively. Rock beats scissors, scissors beats paper and paper beats rock. In all other cases the players tie. If one player beats the other, the winning player receives one dollar from the other player. The payoff matrix corresponding to this game is shown below:

Row\Column	Rock	Paper	Scissors
Rock	0	−1	1
Paper	1	0	−1
Scissors	−1	1	0

Many will know, from experience, that consistently favoring one strategy over the other two is never a good idea. One should 'mix' among them. We model this 'mixing' by allowing players to randomize over their pure strategies. A **mixed strategy** is a probability distribution over the set of pure strategies. The mini–max theorem of zero-sum games, which we prove here, formalizes the intuition that playing a mixed strategy is a good idea.

What strategies should a player choose? Suppose the Row player makes a prediction about what the Column player will do (prediction means a probability distribution over Column's pure strategies). A prediction is a vector $y \in \mathbb{R}^n$, were y_j is the probability that Column chooses column j. Thus $y_j \geq 0$ and $\sum y_j = 1$. Since the Row player is 'risk neutral' she will pick the strategy that maximizes her expected payoff:

$$\arg \max_i \sum_{j=1}^{n} (a_{ij} y_j).$$

Suppose the Row player is pessimistic; she believes that whatever she does Column will play in such a way as to minimize her payoff. Equivalently, Column will choose the vector y so as to minimize $\max_i \sum_{j=1}^{n} (a_{ij} y_j)$. Therefore, to identify this pessimistic choice one needs to solve the following optimization problem:

$$\min \left[\max_i \sum_{j=1}^{n} (a_{ij} y_j) \right]$$

$$\sum_{j=1}^{n} y_j = 1, \quad y_j \geq 0.$$

This is not a linear program but can be transformed into one (which we call LPC)

$$\min R \quad \text{(the mini–max value)}$$

$$\text{s.t.} \sum_{j=1}^{n} (a_{ij} y_j) \leq R,$$

$$\sum_{j=1}^{n} y_j = 1, \quad y_j \geq 0.$$

Now, if we switch Row with Column in the above definitions, the Column player chooses a pure strategy from the set $\arg \min_j \sum_{i=1}^{m} (a_{ij} x_i)$. From Column's point of view, the pessimistic prediction of what Row will do is found by solving

$$\max_x \left[\min_j \sum_{i=1}^{m} (a_{ij} x_i) \right].$$

This can be formulated as an LP (called LPR)

$$\max C$$

$$\text{s.t. } \sum_{i=1}^{m}(a_{ij}x_i) \geq C$$

$$\sum_{i=1}^{m} x_i = 1, \quad x_i \geq 0.$$

The above linear programs are each other's duals.

$\min R$	$\max C$
$x_i \rightarrow R - \sum_{j=1}^{n}(a_{ij}y_j)$	$\sum_{i=1}^{m} x_i = 1$
$C \rightarrow \sum_{j=1}^{n} y_j = 1, \quad y_j \geq 0$	$-\sum_{i=1}^{m}(a_{ij}x_i) + C \leq 0, \quad x_i \geq 0.$

Since both programs are feasible, the duality theorem applies, i.e., $R = C$. What one expects to win, the other expects to loose. So, if both players are pessimistic, they will play in a way as to confirm each others beliefs.

A different story now. The Row player, will decide on a randomized strategy, and inform the Column player of that choice. The Column player will choose her mixed strategy to minimize Row's payoff.

In this context a randomized strategy is simply a probability vector of rows and columns. Let $\Pi(R) = \{x \in \mathbb{R}^m: \sum_i = 1, x \geq 0\}$ and $\Pi(C) = \{y \in \mathbb{R}^n: \sum_j y_j = 1, y \geq 0\}$ be the space of mixed strategies for Row and Column respectively.

The expected payoff to Row, if she chooses mixed strategy x and Column chooses mixed strategy y is:

$$xAy = \sum_{j=1}^{n}\sum_{i=1}^{m}(a_{ij}x_i y_j).$$

Definition 4.13 *A pair of mixed strategies x^*, y^* are an* **equilibrium** *if they satisfy:*

$$x^*Ay^* \geq xAy^*, \quad \forall x \in \Pi(R)$$

and

$$x^*Ay^* \leq x^*Ay, \quad \forall y \in \Pi(C).$$

Theorem 4.14   *Let x^* be the optimal solution to LPR and y^* the optimal solution to LPC. Then (x^*, y^*) is an equilibrium.*

Proof   For the pair x^*, y^*, the expected payoff to row is

$$x^* A y^* = \sum_{j=1}^{n} \sum_{i=1}^{m} (a_{ij} x_i^* y_j^*) = \sum_{i=1}^{m} x_i^* \left(\sum_{j=1}^{n} (a_{ij} y_j^*) \right).$$

By complementary slackness

$$\sum_{i=1}^{m} x_i^* \left(R - \sum_{j=1}^{n} (a_{ij} y_j^*) \right) = 0,$$

i.e., $\sum_{i=1}^{m} x_i^* \left(\sum_{j=1}^{n} (a_{ij} y_j^*) \right) = R$. For any randomized strategy $x^0 \neq x^*$,

$$\sum_{i=1}^{m} x_i^0 \left(\sum_{j=1}^{n} (a_{ij} y_j^*) \right) \leq \sum_{i=1}^{m} (x_i^0 R) = R = x^* A y^*. \qquad \blacksquare$$

4.7   Application: Afriat's theorem

Sydney Afriat's theorem is an answer to the question of when a sequence of purchase decisions is consistent with the purchaser maximizing a concave utility function $u(\cdot)$.[10]

Imagine a purchaser contemplating how much of each of n goods should be purchased. The quantity can be represented by a vector $x \in \mathbb{R}_+^n$. The price of each good can be represented by a vector $p \in \mathbb{R}_+^n$. Suppose a sequence of purchase decisions (p^i, x^i), $i = 1, \ldots, n$, where $p^i \in \mathbb{R}_+^n$ is the price vector and $x^i \in \mathbb{R}_+^n$ the corresponding purchased quantity. Suppose the purchaser makes her purchase decisions based on utility maximization.

If $p^i \cdot (x^j - x^i) \leq 0$, it means that at the vector of prices p^i, bundle x^i is at least as expensive as bundle x^j. We know that the purchaser chose bundle x^i, the more expensive bundle, over x^j. Thus she must assign more utility to bundle x^i than to bundle x^j. Therefore the utility function u must satisfy $u(x^j) \leq u(x^i)$.

If we have a sequence of decisions $(p^{i_1}, x^{i_2}), (p^{i_2}, x^{i_2}), (p^{i_3}, x^{i_3}), \ldots, (p^{i_k}, x^{i_k})$, with

$$p^{i_1} \cdot (x^{i_2} - x^{i_1}) \leq 0, \quad p^{i_2} \cdot (x^{i_3} - x^{i_2}) \leq 0, \quad \ldots, \quad p^{i_k} \cdot (x^{i_1} - x^{i_k}) \leq 0,$$

we must by the same reasoning conclude that $u(x^{i_1}) \leq u(x^{i_2}) \leq \cdots \leq u(x^{i_k}) = u(x^{i_1})$, i.e., $u(x^{i_1}) = u(x^{i_2}) = \cdots = u(x^{i_k})$. Since all the bundles in this sequence have the same utility, they must cost the same, i.e.,

$$p^i \cdot (x^j - x^i) = 0, \quad p^j \cdot (x^k - x^j) = 0, \quad \ldots, \quad p^r \cdot (x^i - x^r) = 0.$$

The above *necessary* condition can be described in graph theoretic terms. Let A be a $n \times n$ matrix of real numbers with all zero's on the diagonals. Let $a_{ij} = p^i \cdot (x^j - x^i)$ for all $i \neq j$. We associate with the matrix A a directed graph $D(A)$ as follows: introduce a vertex for each index and for each ordered pair (i, j) an edge with length a_{ij}. The matrix A will be said to satisfy the Afriat condition (AC) if every negative length cycle in $D(A)$ contains at least one edge of positive weight.

Associated with A is an inequality system

$$y_j \leq y_i + s_i a_{ij}, \quad \forall i \neq j, \ 1 \leq i, j \leq n,$$
$$s_i > 0, \quad \forall 1 \leq i \leq n.$$

We label it $L(A)$. We now state Afriat's theorem.

Theorem 4.15   *$L(A)$ is feasible iff $D(A)$ satisfies AC.*

Whenever $D(A)$ satisfies AC, we use the solution to $L(A)$ to construct a concave utility function $u(\cdot)$ consistent with the sequence of purchase decisions (p^i, x^i) by setting

$$u(x) = \min\{y_1 + s_1 p^1(x - x^1), y_2 + s_2 p^2(x - x^2), \ldots, s_n p^n(x - x^n)\}.$$

We will use the duality theorem to prove Afriat's theorem.[11]
Consider the following linear program:

$$\min \ 0 \cdot s + 0 \cdot y$$
$$\text{s.t.} \quad s_i \geq 1, \quad \forall i,$$
$$a_{ij} s_i + y_i - y_j \geq 0, \quad \forall i \neq j.$$

Feasibility of this program yields Afriat's theorem. We will show that its dual is feasible and has objective function value zero, from which it will follow that the primal is feasible.

Let z_i be the dual variable associated with the constraint $s_i \geq 1$ and x_{ij} the dual variable associated with the constraint $a_{ij} s_i + y_i - y_j \geq 0$. The dual is

$$\max \sum_i z_i$$
$$\text{s.t.} \quad \sum_k x_{ki} - \sum_j x_{ij} = 0, \quad \forall i,$$

$$z_i + \sum_j a_{ij}x_{ij} = 0, \quad \forall i,$$

$$z_i, x_{ij} \ge 0, \quad \forall i, j.$$

With every solution (x, z) of this linear program we can associate a directed graph as follows: one vertex for each index i and an arc directed from i to j if $x_{ij} > 0$. Call this directed graph $D(x)$. An arc (i, j) of $D(x)$ will be called **non-singular** if $a_{ij} \ne 0$.

Lemma 4.16 *There is an optimal solution, (x^*, z^*), to the dual LP such that every cycle in $D(x^*)$ contains a non-singular arc.*

Proof   Suppose (x, z) to be an optimal solution and C a cycle in $D(x)$ with no non-singular arc. Construct a new solution (x', z') as follows:

1. $x'_{ij} = x_{i,j} \; \forall (i, j) \notin C$
2. $x'_{ij} = x_{i,j} - \epsilon \forall (i, j) \in C$ for $\epsilon > 0$ sufficiently small.
3. $z' = z$.

Since $x_{ij} > 0$ for all (i, j), we can choose $\epsilon > 0$ sufficiently small so that $x'_{ij} \ge 0$ for all (i, j). In fact, choose ϵ to make at least one of x'_{ij} for $(i, j) \in C$ equal to zero. We now show that (x', z') is dual feasible.

Since $a_{ij} = 0$ for all $(i, j) \in C$ it follows that for all i,

$$z'_i + \sum_j a_{ij}x'_{ij} = z_i + \sum_{j:(i,j)\in C} 0 \times x'_{ij} + \sum_{j:(i,j)\notin C} x'_{ij} = z_i + \sum_j a_{ij}x_{ij} = 0.$$

Next, consider the term $\sum_k x'_{ki} - \sum_j x'_{ij}$ for each i. Suppose first that there is no index q such that (q, i) or (i, q) is in C. Then $x'_{ki} = x_{ki}$ and $x'_{ij} = x_{ij}$ for all $k, j \ne i$. Thus $\sum_k x'_{ki} - \sum_j x'_{ij} = \sum_k x_{ki} - \sum_j x_{ij} = 0$. Now suppose there is such an index. Since C is a cycle there must be exactly two indices k', j' such that $(k', i), (i, j') \in C$. In this case

$$\sum_k x'_{ki} - \sum_j x'_{ij} = x'_{k'i} + \sum_{k \ne k'} x_{ki} - \sum_{j \ne j'} x_{ij} - x'_{ij'}$$

$$= x_{k'i} - \epsilon + \sum_{k \ne k'} x_{ki} - \sum_{j \ne j'} x_{ij} - x_{ij'} + \epsilon$$

$$= \sum_k x_{ki} - \sum_j x_{ij} = 0.$$

Thus (x', z') is feasible and has the same objective function value as (x, z). Finally, C is not a cycle present in $D(x')$. Now repeat the argument with $D(x)$ replaced by $D(x')$. ∎

Theorem 4.17 *There is an optimal solution to the dual program, (x^*, z^*), such that $x^* = 0$ and $z^* = 0$.*

Proof Suppose not. Choose (x^*, z^*) to satisfy the conditions of the previous lemma. Then there is an index i_1 such that $z_{i_1}^* > 0$, i.e., $\sum_k a_{i_1 k} x_{i_1 k}^* < 0$. Thus there is an index, i_2 such that $a_{i_1 i_2} x_{i_1 i_2} < 0$, i.e., $a_{i_1 i_2} < 0$ and $x_{i_1 i_2} > 0$. From this we deduce that $\sum_k x_{i_2 k} > 0$ and $\sum_{k:x_{i_2 k}>0} a_{i_2 k} x_{i_2 k} \leq 0$. So, there is an index i_3, say, such that $x_{i_2 i_3} > 0$ and $a_{i_2 i_3} \leq 0$. Now repeat the argument. Since the number of indices is finite we must eventually repeat an index, i.e., we have identified a cycle, C, in $D(x^*)$. By construction $a_{ij} \leq 0$ for every arc $(i, j) \in C$. By AC, $a_{ij} = 0$ for all $(i, j) \in C$ which violates our choice of (x^*, z^*). ∎

4.8 Integer programming

An **integer program** is a linear program with the additional requirement that the solution be integral. A full discussion of integer programming deserves a book by itself.[12] Here we limit ourselves to describing one sufficient condition that guarantees that the extreme points of a linear program are integral.

Before proceeding you should convince yourself that no 'simple' scheme based on solving the underlying linear program and rounding the resulting solution can find the optimal integer solution.

Definition 4.18 *A matrix is called* **totally unimodular** *(TUM) iff the determinant of each of its square submatrices has value 1, −1 or 0.*

If a matrix is TUM then so is its transpose. If A and E are TUM, then so is AE.

Example 16 *The following matrix:*

$$\begin{bmatrix} 1 & 0 \\ 0 & 1 \end{bmatrix}$$

is obviously TUM. The following is not:

$$\begin{bmatrix} 1 & 1 & 0 \\ 0 & 1 & 1 \\ 1 & 0 & 1 \end{bmatrix}.$$

Every proper square submatrix has a determinant with absolute value 0 or 1. However the determinant of the entire matrix is 2.

Theorem 4.19   *Let A be a m × n TUM matrix all of whose entries are integral. Let b be a m × 1 integral vector. Then every extreme point of $\{Ax = b, x \geq 0\}$ is integral.*

Proof   To every extreme point w of $\{Ax = b, x \geq 0\}$ there is a basis of A such that $w = B^{-1}b$. By Cramer's rule, we can write $B^{-1} = B^*/\det B$ where B^* is the adjoint of B. Since A has all integral entries, B^* has all integer entries. Since A is TUM and B is non-singular, it follows that $|\det B| = 1$. Hence B^{-1} has all integer entries. Thus $B^{-1}b$ is integral.   ■

For most applications, the following characterization of TUM matrices for restricted classes of matrices is the tool of choice.

Theorem 4.20   *Let A be a matrix each of whose entries is 0, 1 or −1. Suppose each subset S of columns of A can be divided into two sets L and R such that*

$$\left| \sum_{j \in S \cap L} a_{ij} - \sum_{j \in S \cap R} a_{ij} \right| = 0, 1, \quad \forall i.$$

Then A is TUM and the converse is also true.

Proof   First assume that A is TUM. Fix a subset S of columns and define the vector z by $z_j = 1$ if $j \in S$ and zero otherwise. Set $b = Az$. Define vectors l and u as follows:

1. $l_i = \frac{b_i}{2}$ if b_i is even,
2. $l_i = \frac{b_i}{2} - \frac{1}{2}$ if b_i is odd,
3. $u_i = \frac{b_i}{2}$ if b_i is even,
4. $u_i = \frac{b_i}{2} + \frac{1}{2}$ if b_i is odd.

Consider the polyhedron:

$$P = \{x \in \mathbb{R}_+^m : l \leq Ax \leq u, x \leq z\}.$$

Since $z/2 \in P$, the polyhedron is non-empty and must have at least one extreme point. Since A is TUM, the extreme point must be integral, in fact, since $0 \leq x \leq z$ it is 0–1. Call the extreme point x^*. Observe that $x_j^* = z_j$ if $j \in S$ and $x_j^* = 0$ if $j \notin S$. Hence $z_j - 2x_j^*$ is either 1 or −1 for all $j \in S$. Set $L = \{j \in J : z_j - 2x_j^* = 1\}$ and $R = \{j \in J : z_j - 2x_j = -1\}$.
Then

$$\sum_{j \in L} a_{ij} - \sum_{j \in R} a_{ij} = \sum_{j \in J} a_{ij}(z_j - 2x_j^*).$$

Given $l_i \le \sum_j a_{ij} x^* j \le u_i$ it follows that the right-hand side of the above sum is either 0, 1 or -1.

Now suppose the partition property holds, We show that A is TUM. The proof will be by induction on the size of square submatrices of A. Choose $S = \{j\}$. From the partition property, a_{ij} will be 0, 1 or -1. This shows the induction hypothesis for all square sub-matrices of size 1. Now suppose true for all $(k-1) \times (k-1)$ submatrices. Let B be any $k \times k$ non-singular submatrix of A. Set $\Delta = |\det B|$.

By the induction hypothesis, each entry of the adjoint of B, B^* will be 0, 1 or -1. Since $B^{-1} = B^*/\Delta$ it follows that $Bb^*(1) = \Delta e^1$ where $b^*(1)$ is the first column of B^* and e^1 is the vector whose first component is 1 and all others are zero.

Set $J = \{i: b_{i1}^* \ne 0\}$ and $J_1' = \{i: b_{i1}^* = 1\}$. Observe that $J \ne \varnothing$. From $Bb^*(1) = \Delta e^1$, we have for $i = 2, \ldots, k$ that

$$\sum_{j \in J_1'} b_{ij} - \sum_{j \in J \setminus J_1'} b_{ij} = 0.$$

Thus $|\{i \in J: b_{ij} \ne 0\}|$ is even. Hence, for any partition L and R of J it follows that $\sum_{j \in L} b_{ij} - \sum_{j \in R} b_{ij}$ is even for $i = 2, \ldots, k$ because

$$\sum_{j \in L} b_{ij} - \sum_{j \in R} b_{ij} = \sum_{j \in J} b_{ij} - 2 \sum_{j \in R} b_{ij}.$$

By the partition assumption we can choose L and R so that

$$\left| \sum_{j \in L} b_{ij} - \sum_{j \in R} b_{ij} \right| \le 1.$$

Hence

$$\left| \sum_{j \in L} b_{ij} - \sum_{j \in R} b_{ij} \right| = 0, \quad \forall i = 2, \ldots, k.$$

Consider now $t = |\sum_{j \in L} b_{1j} - \sum_{j \in R} b_{1j}|$. Suppose first that $t = 0$. Define the vector z by $z_j = 1$ if $j \in L$, $z_j = -1$ if $j \in R$ and zero otherwise. Because $t = 0$ we have that

$$\left| \sum_{j \in L} b_{ij} - \sum_{j \in R} b_{ij} \right| = 0, \quad \forall i = 1, 2, \ldots, k.$$

Hence $Bz = 0$. Since B is non-singular this implies that $z = 0$ and so $J = L \cup R = \varnothing$ a contradiction. So, we conclude that $t = 1$. Thus Bz is either e^1

or $-e^1$. Suppose the first (a similar argument applies to the second possibility). Then $Bb^*(1) = \Delta Bz$, i.e., $b^*(1) = \Delta z$. However both $b^*(1)$ and z are vectors all of whose entries are 0, 1 or -1. Hence $|\Delta| = 1$. ∎

A simple consequence of the above theorem is that a matrix satisfying all of the following conditions is TUM:

1. Each entry is 0, 1 or -1.
2. Each row contains at most two non-zero entries.
3. The entries are of opposite sign.

Since the property of being TUM is preserved under transposition, we can replace 'row' by 'column' in the above list.

Example 17 *The following matrix is TUM:*

$$\begin{bmatrix} 1 & 1 & 0 & 0 \\ 0 & 1 & 1 & 0 \\ 0 & 0 & 1 & 1 \\ 1 & 0 & 0 & 1 \end{bmatrix}.$$

To see why, multiply the first and second columns by -1. Notice that this preserves the property of TUM. The resulting matrix is:

$$\begin{bmatrix} -1 & 1 & 0 & 0 \\ 0 & 1 & -1 & 0 \\ 0 & 0 & -1 & 1 \\ -1 & 0 & 0 & 1 \end{bmatrix}.$$

Notice that it contains at most two non-zero entries in each row and they are of opposite sign.

4.9 Application: efficient assignment

Consider an economy where the set of agents is denoted N and consisting of a set of indivisible (not necessarily identical) objects denoted M. Let $v_{ij} \geq 0$ be the monetary value that agent $j \in N$ assigns to object $i \in M$. Each agent is interested in consuming at most one object. By adding dummy objects of zero value to all agents, we can always ensure that $|M| \geq |N|$.

An **assignment** of the objects to agents, is an allocation of objects to agents so that no agent receives more than one object and no object is assigned to more than one agent. An **efficient assignment** is one that maximizes the sum of valuations of the agents. The problem of finding the efficient allocation is sometimes called the social planner's problem.

To formulate the problem of finding an efficient assignment as an integer program let $x_{ij} = 1$ if agent j is allocated object i and zero otherwise.

$$\max \sum_{j \in N} \sum_{i \in M} v_{ij} x_{ij}$$

$$\text{s.t. } \sum_{j \in N} x_{ij} \le 1, \quad \forall i \in M,$$

$$\sum_{i \in M} x_{ij} \le 1, \quad \forall j \in N,$$

$$x_{ij} \in \{0, 1\}, \quad \forall i \in M, \ j \in N.$$

This problem is an instance of the **assignment problem**. First we show that the polyhedron obtained by dropping the integrality restriction has integral extreme points. That polyhedron is described:

$$\text{s.t. } \sum_{j \in N} x_{ij} \le 1, \quad \forall i \in M,$$

$$\sum_{i \in M} x_{ij} \le 1, \quad \forall j \in N,$$

$$x_{ij} \ge 0, \quad \forall i \in M, \ j \in N.$$

The first two constraints ensure that $x_{ij} \le 1$ for all i and j.

We show that the constraint matrix of this system is TUM. Fix a good i and agent j. Consider the column associated with the variable x_{ij}. The variable appears with a coefficient of 1 in exactly two rows. One occurs in a row corresponding to agent j and the other to a row corresponding to object i. Let L consist of all rows corresponding to objects and R the set of all rows corresponding to agents. Multiply all the rows in L by -1. We now have a constraint matrix where each column contains exactly two non-zero entries of opposite sign. Given the TUM property the problem of finding the efficient assignment reduces to the following linear program:

$$\max \sum_{j \in N} \sum_{i \in M} v_{ij} x_{ij}$$

$$\text{s.t. } \sum_{j \in N} x_{ij} \le 1, \quad \forall i \in M,$$

$$\sum_{i \in M} x_{ij} \le 1, \quad \forall j \in N,$$

$$x_{ij} \ge 0, \quad \forall i \in M, \ j \in N.$$

Let p_i be the dual variable associated with each constraint $\sum_{j \in N} x_{ij} \le 1$ and λ_j the dual variable associated with each constraint $\sum_{i \in M} x_{ij} \le 1$. The dual to the above program is:

$$\min \sum_{j \in N} \lambda_j + \sum_{i \in M} p_i$$

$$\text{s.t.} \ \lambda_j + p_i \ge v_{ij}, \quad \forall j \in N, \ \forall i \in M,$$

$$\lambda_j, p_i \ge 0, \quad \forall j \in N, \ \forall i \in M.$$

The dual has an interesting interpretation. One can think of each p_i as the price of object i. Given a collection of prices, the optimal solution to the dual is found by setting each λ_j to $\max_{i \in M}(v_{ij} - p_i)$. Thus, each λ_j represents the maximum surplus that agent j can receive from the consumption of a single object at prices $\{p_i\}_{i \in M}$.

Suppose x^* is an optimal integral solution to the primal and (λ^*, p^*) an optimal solution to the dual. Then the prices p^* 'support' the efficient assignment x^* in the following sense. Suppose we post a price p_i^* for each $i \in M$. Next, ask each agent to name the set of objects that maximize their surplus at the posted prices. Then, it is possible to give each agent exactly one of their named objects. To see why this last statement must be true, recall complementary slackness

$$(\lambda_j^* + p_i^* - v_{ij})x_{ij}^* = 0.$$

So, if $x_{ij}^* = 1$ it follows that $\lambda_j^* = v_{ij} - p_i^* = \max_{r \in M}(v_{rj} - p_r^*)$. Hence, in this economy there is a set of prices that can be posted for each good so as to balance supply with demand.

One can associate with the problem of finding an efficient assignment a cooperative game with a non-empty core. This is discussed in Section 7.4.

4.10 Application: Arrow's theorem

Kenneth Arrow's impossibility theorem is the most famous theorem of economic theory. First, it establishes the impossibility of aggregating diverse preferences in some 'democratic' way. Second, it has provided work for many a theorist.[13]

Arrow's set up has a collection of voters each with a strict preference ordering over a finite set A of alternatives (at least two).[14] The goal is to identify a single preference ordering over A, that in some sense best reflects the disparate orderings of the voters. Just as the mean of a set of numbers summarizes those numbers, so we seek a preference ordering that summarizes the different orderings of the agents.

One can imagine a variety of schemes for summarizing a collection of preference orderings. Rather than build a long list of such schemes and compare and contrast them, Arrow argued that one should identify attractive properties or axioms that such schemes should satisfy and then deduce which schemes possessed them.

Arrow advanced two axioms and showed that only one, rather unpalatable scheme satisfied both.

Let Ω denote the set of all strict preference orderings over the alternatives in A.[15] Let Ω^n be the set of all n-tuples of preferences from Ω. An element of Ω^n will be denoted $\mathbf{P} = (\mathbf{p}_1, \mathbf{p}_2, \ldots, \mathbf{p}_n)$ and called a **profile**. One should think of a profile as a list of preference orderings, one for each voter. An n-person **social welfare function** (SWF) on Ω is a function $f: \Omega^n \to \Omega$.

Three points deserve comment. First, we are interested in a function that will take as input a profile of preference orderings and return an ordering rather than a single alternative or subset of alternatives. There is a strand of the literature that considers these other possibilities. Second, the function must return an ordering for every possible profile of orderings. Another strand of the literature examines the consequences of relaxing this requirement on the grounds that in some contexts, certain profiles of preferences are simply inconceivable. Third, only information about orderings is used, 'intensity' of preference is ignored.

One example of a SWF is the **dictatorial** one. Fix an agent i and set $f(\mathbf{P}) = \mathbf{p}_i$ for all $\mathbf{P} \in \Omega$. As the label suggests, this SWF summarizes the profile by simply picking the preference ordering of a particular agent.

Arrow imposed two conditions on social welfare functions:

1. **Unanimity** (U) If for every $\mathbf{P} \in \Omega^n$ and some $x, y \in A$ we have $x\mathbf{p}_i y$ for all i then $xf(\mathbf{P})y$.
2. **Independence of Irrelevant Alternatives** (IIA) For any $x, y \in A$ suppose that exists $\mathbf{P}, \mathbf{Q} \in \Omega^n$ such that $x\mathbf{p}_i y$ if an only if $x\mathbf{q}_i y$ for $i = 1, \ldots, n$. Then $xf(\mathbf{P})y$ if an only if $xf(\mathbf{Q})y$.

The first is a minimal requirement for any SWF that would claim to summarize a profile that no one can object to. The second says that only pairwise comparisons matter. Thus when a SWF must decide on whether to rank x above y or vice-versa, only the pattern of preferences with respect to x and y matter. Whether chicken is to be ranked above or below beef should not depend on the presence or absence of some third alternative, fish, say. The second is not so benign and the reader must look elsewhere for the arguments for and against.

Arrow's theorem states that the only SWF on Ω^n that satisfies U and IIA is the dictatorial one. To save on notation we consider only 2-person SWFs. The arguments can be extended to the case of n-person SWFs.

Fix an SWF, f that satisfies IIA. Consider a profile $\mathbf{P}$ and a pair $x, y \in A$. Suppose agent 1 ranks x over y in the profile while agent 2 ranks y above x. If $xf(\mathbf{P})y$ then IIA implies that whenever agent 1 ranks x over y and agent 2 the reverse, the SWF f will rank x above y. In this case we say that agent 1 is **decisive** over the ordered pair (x, y). The observation allows us to describe a SWF that satisfies IIA in terms of which ordered pair of alternatives agent 1 is decisive over.

Denote the set of all ordered pairs of alternatives by A^2. For each element $(x, y) \in A^2$ we define a $0-1$ variable as follows:

- $d(x, y) = 1$ if agent 1 is decisive for x over y,
- $d(x, y) = 0$ otherwise.

If $d(x, y) = 0$ it is to be understood that agent 2 will be decisive for x over y.

To ensure that the assignment implied by the d variables satisfies unanimity as well as produces an ordering of the alternatives we need to impose additional conditions. They are described below.

Suppose there are $\mathbf{p}, \mathbf{q} \in \Omega$ and three alternatives x, y and z such that $x\mathbf{p}y\mathbf{p}z$ and $y\mathbf{q}z\mathbf{q}x$. Since Ω is the set of all possible orderings a pair like $\mathbf{p}$ and $\mathbf{q}$ exist in Ω. Suppose agent 1 has the ordering $\mathbf{p}$ and agent 2 has the ordering $\mathbf{q}$. Suppose $d(x, y) = 1$. Since agent 1 ranks x over y and agent 2 the reverse, the SWF ranks x above y. Both agents rank y above z, so by U the SWF must rank y above z. To ensure that an ordering is produced the SWF must rank x over z. Notice, however, that agent 1 is the only person who ranks x above z. Thus requiring $d(x, y) = 1$ forced us to set $d(x, z) = 1$. To summarize,

$$d(x, y) = 1 \Rightarrow d(x, z) = 1 \quad \text{and}$$
$$d(z, x) = 1 \Rightarrow d(y, x) = 1.$$

The last implication is derived by endowing agent 1 with the ordering $\mathbf{q}$ and agent 2 with the ordering $\mathbf{p}$.

These two logical conditions can be formulated as inequalities

$$d(x, y) \leq d(x, z) \quad \text{and} \quad d(z, x) \leq d(y, x). \tag{4.1}$$

Every 2-person ASWF corresponds to a feasible $0-1$ solution to the system (4.1), but not the reverse. The constraint matrix of the system is TUM. Two obviously feasible solutions are $d(x, y) = 1, \forall (x, y) \in A^2$ and $d(x, y) = 0$ for all $(x, y) \in A^2$. The first corresponds to an ASWF where agent 1 is the dictator and the second, by default, to where agent 2 is the dictator. We refer to these solutions as the all 1's and all 0's solution respectively. We show that these are the only feasible $0-1$ solutions.

With each ordered pair of alternatives we associate a vertex. If there is an inequality of the form $d(a, b) \leq d(x, y)$ where (a, b) and (x, y) are ordered pairs of alternatives insert a *directed* edge from (a, b) to (x, y). Call the resulting directed graph D^Ω.

Now we need only verify that between any ordered pair of vertices of D^Ω there is a directed path from one to the other. So, when $d(x, y)$ is set to 1 for any ordered pair (x, y), $d(u, v)$ is set to 1 for all ordered pairs (u, v).

Each vertex corresponds to an ordered pair. Let the pairs corresponding to these two vertices be (x, y) and (u, v). Since Ω contains all possible orderings, the following triples are possible: $\{x, y, u\}$, $\{x, y, v\}$, $\{x, u, v\}$, $\{y, u, v\}$. In particular from inequalities of type (4.1) we get $d(x, y) \leq d(x, u)$, $d(x, u) \leq d(v, u)$ and so there is a path $(x, y) \to (x, u) \to (u, v)$. To see how this is possible consider the following list of orderings: $x\mathbf{p}^1 y\mathbf{p}^1 z$, $y\mathbf{p}^2 z\mathbf{p}^2 x$ and $z\mathbf{p}^3 x\mathbf{p}^3 y$. Applying the inequalities of type (4.1) to $\mathbf{p}^1$ and $\mathbf{p}^2$ yields $d(x, y) \leq d(x, z)$ and $d(z, x) \leq d(y, x)$. Applying the type (4.1) inequalities to $\mathbf{p}^2$ and $\mathbf{p}^3$ yields $d(y, z) \leq d(y, x)$ and $d(x, y) \leq d(z, y)$. Now use $\mathbf{p}^3$ and $\mathbf{p}^1$ to generate $d(z, x) \leq d(z, y)$ and $d(y, z) \leq d(x, z)$. Now consider the orderings $z\mathbf{q}^1 y\mathbf{q}^1 x$, $y\mathbf{q}^2 x\mathbf{q}^2 z$ and $x\mathbf{q}^3 z\mathbf{q}^3 y$. Orderings $\mathbf{q}^1$ and $\mathbf{q}^2$ produce $d(z, y) \leq d(z, x)$ and $d(x, z) \leq d(y, z)$. The pair $\mathbf{q}^2$ and $\mathbf{q}^3$ give us $d(y, x) \leq d(y, z)$ and $d(z, y) \leq d(x, y)$. Finally, the pair $\mathbf{q}^3$ and $\mathbf{q}^1$ yield $d(x, z) \leq d(x, y)$ and $d(y, x) \leq d(z, x)$. Combining them all together produces:

$$d(x, y) \leq d(x, z) \leq d(y, z) \leq d(y, x) \leq d(z, x) \leq d(z, y) \leq d(x, y).$$

Hence, they are either all '1' or all 'zero'. The graph for this system of vertices is shown in Figure 4.4.

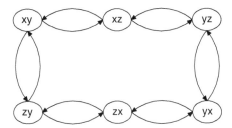

Figure 4.4

Problems

4.1 Compute all basic solutions to the system

$$x_1 - x_2 - x_3 = 0,$$
$$x_1 + 2x_2 - 3x_3 = 1.$$

4.2 Write down the dual to the following linear program:

$$\begin{aligned}
\max \quad & x_1 + 2x_2 \\
\text{s.t.} \quad & x_1 + \tfrac{8}{3}x_2 \leq 4, \\
& x_1 + x_2 = 2, \\
& 2x_1 \geq 3, \\
& x_1 \geq 0.
\end{aligned}$$

4.3 Find the optimal primal and dual solutions to the following LP:

$$
\begin{array}{llll}
\min & x_1 + x_2 & - & 3x_3 \\
\text{s.t.} & x_1 + 2x_2 & - & 3x_3 & = & 4, \\
& 4x_1 + 5x_2 & - & 9x_3 & = & 13, \\
& & & x_1, x_2, x_3 \geq & 0.
\end{array}
$$

4.4 Consider the following LP:

$$
\begin{array}{lllll}
\min & -x_1 + 2x_2 & + 8x_3 + & & 2x_4 \\
\text{s.t.} & x_3 + & x_4 & & \geq x_2 + 1, \\
& x_1 + 2x_2 & - 2x_3 + & & x_4 \leq 2, \\
& & & x_1, x_2, x_3, x_4 \geq & 0.
\end{array}
$$

Find its optimal primal and dual solutions.

4.5 Exhibit an example of a linear program such that it and its dual is infeasible.

4.6 Show how the duality theorem of linear programming can be used to prove the Farkas lemma.

4.7 Write down the duals to the following LPs:

$$\max\{cx: Ax = b, x \geq 0\},$$

$$\min\{cx: Ax = b, x \geq 0\},$$

$$\max\{cx: Ax \leq b, x \geq 0\},$$

$$\min\{cx: Ax \geq b, x \geq 0\}.$$

4.8 Convert the following optimization problem into a single linear program:

$$\min\ |x| + |y| + |z|$$

$$\text{s.t.}\quad x + y \leq 1,$$

$$2x + z = 3.$$

4.9 Let

$$Z = \max \sum_{j=1}^{n} c_j x_j$$

$$\text{s.t.} \sum_{j=1}^{n} a_j x_j \leq b,$$

$$x_j \geq 0, \quad \forall j.$$

Assume that $\{c_j\}_{j\geq 1}, \{a_j\}_{j\geq 1}$ are all positive. Show that $Z = b \max_j c_j/a_j$.

4.10 Consider the LP $\max\{cx: Ax \leq b, x \geq 0\}$. If feasible, show that either it or its dual has an unbounded feasible region.

4.11 (Challenging) Consider the following LP: $\min\{cx: Ax \geq b, x \geq 0\}$. Let x and y be an optimal primal dual pair. Complementary slackness tells us that $x_j(\sum_i a_{ij} y_i - c_j) = 0$ for all j. **Strict** complementary slackness says that for all j either $x_j = 0$ or $\sum_i a_{ij} y_j - c_j = 0$ but not both. Prove that every feasible LP admits an optimal primal-dual pair that satisfies strict complementary slackness.

4.12 Show that for every extreme point of the system $P = \{x : Ax \leq b, x \geq 0\}$, there is a vector c that attains its maximum at that extreme point of P.

4.13 Consider the following **fractional program**:

$$\max \quad \frac{cx + \alpha}{dx + \beta},$$
$$\text{s.t.} \quad Ax \leq b,$$
$$x \geq 0.$$

Here A is an m by n matrix and $x \in \mathbb{R}^n$. Assuming that $dx + \beta > 0$ for all feasible x, show how to solve this problem as single linear program.

4.14 Consider the following linear program:

$$Z = \min\{cx: Ax = b, M \geq x \geq 0\}.$$

Here M is a sufficiently large number such that all solutions, x^* to $Ax = b$ satisfy $x^* \leq M$. Suppose this linear program has an optimal solution. Let $\mu \in \mathbb{R}^m$ and

$$Z_\mu = \min\{cx + \mu(Ax - b): M \geq x \geq 0\}.$$

Prove that $Z = \max_{\mu \in \mathbb{R}^m} Z_\mu$.

4.15 Given an $m \times n$ matrix A (real valued entries) let $G = \text{span}(A)$ and $F = \{y: y = Ax \text{ for some } x \gg 0\}$. Consider the following procedure to decide if $F = G$:

- Choose a $p \in \mathbb{R}^n$ with all positive entries. If $Ap = 0$, then $F = G$, otherwise go to the next step.
- Solve (here μ is a scalar)

$$Z = \min \mu$$
$$\text{s.t.} \quad Ax = 0,$$
$$x + \mu p \geq p,$$
$$\mu \geq 0.$$

If $Z = 0$ then $F = G$ otherwise $F \neq G$.

Prove that the procedure is correct.

4.16   Compute the equilibrium randomized strategies for Row and Column in the following zero-sum game:

$$\begin{bmatrix} 3 & -2 & 1 \\ 1 & 3 & -2 \\ -2 & 1 & 3 \end{bmatrix}.$$

Notes

1   Professor of Management Science at the Fisher School of Business, Ohio State University.
2   Fourier (1827).
3   The (in)justice of this is the subject of occasional dinner table conversation. The reader interested in an account of such matters should look elsewhere.
4   Recall that a^j is the jth column of A.
5   The term 'dual' predates linear programming. Having defined the dual it was natural to ask what one should call the original problem from which the dual was conceived. Dantzig's father, suggested 'primal' as the natural antonym. Like 'dual' it is Greek, and means original or primitive.
6   The Roman god of sleep or dreams.
7   Soma is a narcotic distributed in Aldous Huxley's *Brave New World* that induces euphoria and hallucinations.
8   From the Greek meaning self-love.
9   A web site devoted entirely to the game can be found at www.worldrps.com.
10 See Chapter 5 for a definition of concavity and discussion of utility functions.
11 The proof is based on Fostel *et al.* (2004).
12 Three in fact, Schrijver (2003).
13 Morton Kamien asserts that the importance of a paper should be judged by how much employment it provides for other scholars. By this standard, Arrow's theorem is the most important public works project for economic theorists ever.
14 This section is based on joint work with Jay Sethuraman and Teo Chung Piaw (2003).
15 The assumption that preferences are strict is made for simplicity.

References

Fostel, A., Scarf, H. E. and Todd, M. J.: 2004, Two new proofs of Afriat's theorem, *Economic theory* **24**(1), 10.

Fourier, J.-B. J.: 1827, Analyse des travaux de l'acadamie royale des sciences, pendant l'annee 1824, *Histoire de l'Academie Royale des Sciences de l'Institut de France 7*. Partial English translation in: D.A. Kohler, Translation of a Report by Fourier on his work on Linear Inequalities, Opsearch 10 (1973) 38–42.

Huxley, A.: 1932, *Brave New World*, Chatto & Windus, London.

Schrijver, A.: 2003, *Combinatorial optimization: polyhedra and efficiency*, Algorithms and combinatorics, 24, Springer, Berlin, New York.

Sethuraman, J., Chung Piaw, T. and Vohra, R. V.: 2003, Integer programming and arrovian social welfare functions, *Mathematics of Operations Research* **28**(2), 309.

5 Non-linear programming

In this chapter we consider the problem of optimizing a non-linear function subject to a *finite* collection of non-linear constraints. This is called non-linear programming (NLP). Let $M = \{1, 2, \ldots, m\}$ be an index set. For each $i \in M$ we have a continuous and differentiable function $f^i \colon \mathbb{R}^n \to \mathbb{R}$ that will give rise to a constraint. The objective function will be $f^0 \colon \mathbb{R}^n \to \mathbb{R}$ and is assumed to be continuous and differentiable. The problem (P) we consider is

$$\max\{f^0(x) \colon f^i(x) \geq 0, \ \forall i \in M\}. \tag{P}$$

Any NLP can be transformed into the above form, however unlike the LP case, there are pitfalls for the careless.

For each $x \in \mathbb{R}^n$ let $N_\epsilon(x) = \{y \colon d(x, y) < \epsilon\}$ be an ϵ **neighborhood** around x. Let $F = \{x \in \mathbb{R}^n \colon f^i(x) \geq 0, \forall i \in M\}$ be the feasible set. Notice that $M = \emptyset$ implies that $F = \mathbb{R}^n$.

Definition 5.1 *A point x is called a* **local maximum** *for problem (P) if there exists $\epsilon > 0$ such that $f^0(x) \geq f^0(y)$ for all $y \in N_\epsilon(x) \cap F$. A point x is called a* **global maximum** *if x is an optimal solution to problem (P).*

Every global maximum is a local maximum but not conversely. Later we identify conditions under which a local maximum is a global maximum. It should be obvious how to modify the definitions to define local minima and global minima.

Using the fact that f^0 is continuous and differentiable we can approximate f^0 by its Taylor series expansion

$$f^0(x + \epsilon h) = f^0(x) + \epsilon h \cdot \nabla f^0(x) + r(\epsilon).$$

Here ∇f means the n vector whose ith component is $\frac{\partial f}{\partial x_i}$. $r(\epsilon)$ is the error term which is quadratic in ϵ. For $\epsilon > 0$ sufficiently small, $h \cdot \nabla f^0(x) > 0 \Rightarrow f^0(x + \epsilon h) > f^0(x)$. This is a fact we will make frequent use of.

To motivate the necessary conditions for optimality, suppose $x^* \in F$ a local maximum of f^0. Then, in a sufficiently small neighborhood around x^*, x^* must be a global maximum of f^0. The Taylor series expansion of f^0 allows us to approximate the function by a linear function in a sufficiently small neighborhood around x^*. So, let $\delta > 0$ be small enough and

$$f(x^* + \delta h) \rightarrow f(x^*) + \delta h \cdot \nabla f(x^*).$$

Thus, it seems reasonable to suppose that $h = 0$ must be the optimal solution to the following linear program (here h is the variable):

$$\max \quad f^0(x^*) + \delta h \cdot \nabla f^0(x^*)$$
$$\text{s.t.} \quad f^i(x^*) + \delta h \cdot \nabla f^i(x^*) \geq 0, \quad \forall i \in M.$$

Dropping constant terms from the objective function, we can rewrite this linear program as:

$$\max \quad \delta h \cdot \nabla f^0(x^*)$$
$$\text{s.t.} \quad -\delta h \cdot \nabla f^i(x^*) \leq f^i(x^*), \quad \forall i \in M.$$

Its dual is:

$$\min \quad \sum_{i \in M} f^i(x^*) \mu_i$$
$$\text{s.t.} \quad -\sum_{i \in M} \mu_i \nabla f^i(x^*) - \mu_0 \nabla f^0(x^*) = 0, \quad \forall i \in M.$$

Hence, at a local maximum x^*, if $h = 0$ is the optimal solution to the primal, then $\sum_{i \in M} f^i(x^*)\mu_i = 0$. Since μ_i and $f^i(x^*)$ are non-negative for all i, this last equation implies that $\mu_i f^*(x^*) = 0$ for all $i \in M$. Which is, of course, complementary slackness. Second,

$$\mu_0 \nabla f^0(x^*) + \sum_{i \in M} \mu_i \nabla f^i(x^*) = 0.$$

Summarizing we expect the following:
If x^* is a local maximum of problem (P) then there exist non-negative multipliers $\{\mu_i\}_{i \in M}$ such that

$$\mu_0 \nabla f^0(x^*) + \sum_{i \in M} \mu_i \nabla f^i(x^*) = 0$$

and $\mu_i f^*(x^*) = 0$ for all $i \in M$.
   In what follows we shall see just how true this hypothesis is.

5.1 Necessary conditions for local optimality

Here we identify conditions that all local optima must satisfy. However solutions that are not locally optimal may also satisfy these conditions.

Theorem 5.2 (Unconstrained case) *Suppose in problem (P), $M = \varnothing$. If x^* is a local maximum then $\nabla f^0(x) = 0$.*

Proof Suppose not. Let $h = \nabla f^0(x^*) \neq 0$. Then $h \cdot h > 0$. Hence $f^0(x^* + \epsilon h) > f^0(x^*) + \epsilon h \cdot \nabla f^0(x^*) > f(x^*)$ for all $\epsilon > 0$ sufficiently small. This contradicts local optimality of x^*. ∎

Example 18 *Figure 5.1 shows the graph of a function one variable. The slope of the curve at the point x^* is zero, i.e., the derivative of the function at x^* is zero. However, x^* is neither a local maximum or local minimum.*

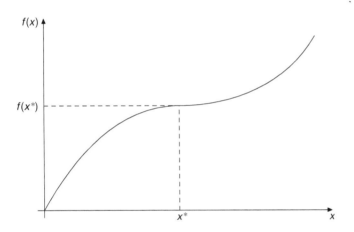

Figure 5.1

Lemma 5.3 (Constrained case) *Suppose $M \neq \varnothing$ and x^* is a local maximum for problem (P). Then there exist non-negative multipliers $\{\mu_0, \mu_1, \ldots, \mu_m\}$ not all zero such that*

$$\mu_0 \nabla f^0(x^*) + \sum_{i \in M} \mu_i \nabla f^i(x^*) = 0. \tag{5.1}$$

Proof Let $z^i = \nabla f^i(x^*)$ for $i \in M \cup \{0\}$. If the lemma is true we can divide equation (5.1) through by $\sum_{i \in M \cup \{0\}} \mu_i$. Equation (5.1) can then be interpreted as saying that the origin is in the convex hull of the z^i's. This is what we will prove. Suppose, for a contradiction that the origin is not in the convex hull of the z^i's. Recall that the convex hull of a finite number of vectors is closed. By the strict separating hyperplane theorem there is a vector h such that $h \cdot z^i > 0$ for $i = 0, 1, \ldots, m$.

Therefore, for all $\epsilon > 0$ sufficiently small $f^0(x^* + \epsilon h) > f^0(x^*)$. Further $x^* + \epsilon h$ is feasible for ϵ sufficiently small because

$$f^i(x^* + \epsilon h) > f^i(x^*) \geq 0, \quad \forall i \in M.$$

This contradicts the local optimality of x^*. ■

The proof above fails if M is an infinite set because we cannot guarantee to find an $\epsilon > 0$ such that $f^i(x^* + \epsilon h) > f^i(x^*)$ for all $i \in M$. If x^* is in the strict interior of the feasible region, the theorem follows from the previous one. So, it is of interest only when x^* is on the boundary of the feasible region.

The necessary condition in the previous theorem tells us nothing about the local maximum if $\mu_0 = 0$. In this case equation (5.1) reduces to

$$\sum_{i \in M} \mu_i \nabla f^i(x^*) = 0,$$

an equation that only contains terms from the constraint set and so valueless from an optimization point of view. Unfortunately there can be NLPs where there is no choice of μ's such that $\mu_0 > 0$.

Example 19 *Let $f^0(x_1, x_2) = x_2$, $f^1(x_1, x_2) = x_1$ and $f^2(x_1, x_2) = -x_1 - x_2^2$. The feasible region is given by*

$$
\begin{aligned}
x_1 &\geq 0, \\
x_1 + x_2^2 &\leq 0.
\end{aligned}
$$

The only feasible solution is $(0, 0)$. Thus, no matter what f^0 is, the only optimal solution is the origin.

Now $\nabla f^0(x) = (0, 1)$, $\nabla f^1(x) = (1, 0)$ and $\nabla f^2(x) = (-1, -2x_2)$. Thus $\nabla f^2(0) = (-1, 0)$. Substituting into (5.1) at the point $x^* = (0, 0)$ yields:

$$\mu_0(0, 1) + \mu_1(1, 0) + \mu_2(-1, 0) = 0.$$

All non-negative solutions to this equation have $\mu_0 = 0$. Figure 5.2 illustrates why $\nabla f^0(0, 0)$ cannot lie in the convex hull $\nabla f^1(0, 0)$ and $\nabla f^2(0, 0)$.

Hence, without additional assumptions we cannot guarantee that $\mu_0 > 0$. These additional assumptions are called **constraint qualifications**. One is described below.

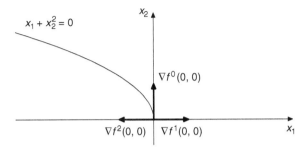

Figure 5.2

Theorem 5.4 (Kuhn–Tucker–Karush theorem) *Suppose $M \neq \varnothing$ and x^* is a local maximum for (P). If the vectors $\{\nabla f^i(x^*)\}_{i \in M}$ are LI then there exist non-negative multipliers $\{\lambda_i\}_{i \in M}$ (not all zero) such that*

$$\nabla f^0(x^*) + \sum_{i \in M} \lambda_i \nabla f^i(x^*) = 0. \tag{5.2}$$

Proof Apply Lemma 5.3. If $\mu_0 > 0$, divide through by μ_0 to obtain the result. If $\mu_0 = 0$ then $\sum_{i \in M} \mu_i \nabla f^i(x^*) = 0$. However LI of the set $\{\nabla f^i(x^*)\}_{i \in M}$ implies that $\mu_i = 0 \ \forall i \in M$ a contradiction. ∎

Stated componentwise equation (5.2) reads

$$\frac{\partial f^0}{\partial x_j} + \sum_{i \in M} \lambda_i \frac{\partial f^i}{\partial x_j} = 0, \quad \forall j.$$

The LI condition is a strong requirement. In some cases it is not needed as the next result shows. We can interpret the next theorem as saying that all $f^i(x)$ being linear is a constraint qualification.

Theorem 5.5 *Let A be an $m \times n$ matrix and $b \in \mathbb{R}^m$. Let f^0 be a continuous differentiable function. Let x^* be a local maximum for $\max\{f^0(x): Ax \geq b\}$. Then, there is a non-zero $y \in \mathbb{R}^m_+$ such that $\nabla f^0(x^*) + A^\mathsf{T} y = 0$ and $y(Ax^* - b) = 0$.*

Proof Consider problem (P) where each constraint function, $f^i = \sum_{j=1}^n a_{ij} x_j - b_i$ for $i \in M$. Then ∇f^i will just be the ith row of A. Hence showing that there are non-negative, non-zero multipliers $y \in \mathbb{R}^m$ such that $\nabla f^0(x^*) + A^\mathsf{T} y = 0$ is equivalent to showing that there are non-negative, non-zero multipliers $y \in \mathbb{R}^m$

such that

$$\nabla f^0(x^*) + \sum_{i=1}^{m} \nabla f^i(x^*) y_i = 0.$$

Let $s = Ax^* - b \geq 0$. Without loss of generality suppose that the first r components of s are 0 and the remaining $m - r$ components are strictly positive. That is the first r constraints of $Ax \geq b$ are binding at x^*.

If no constraints were binding we have an interior solution and so are unconstrained.

If we can show that $-\nabla f^0(x^*)$ can be expressed as a non-negative linear combination of the $\nabla f^i(x^*)$ for $i = 1, \ldots, r$ (i.e. the first r constraint functions) we are done.

Suppose not. By the Farkas lemma there is a $h \in \mathbb{R}^n$ such that $-h \cdot \nabla f^0(x^*) < 0$ and $h \cdot \nabla f^i(x^*) \geq 0$ for $i = 1, \ldots, r$. Note that we can say nothing about the sign of $h \cdot \nabla f^i(x^*)$ for $i > r$. It could be positive or negative.

Choose $\delta > 0$ suitably small. Consider $x^* + \delta h$. We show that it is feasible. Observe that

$$A[x^* + \delta h] = b + s + \delta Ah.$$

If we look at any of the first r rows of $b + s + \delta Ah$:

$$b_i + s_i + \delta \sum_{j=1}^{n} a_{ij} h_j = b_i + \delta h \cdot \nabla f^i(x^*) \geq b_i.$$

The last inequality follows from the fact that $s_i = 0$ and $h \cdot \nabla f^i(x^*) \geq 0$ for $i = 1, \ldots, r$. For rows $r + 1$ and larger, $s_i > 0$. So for $\delta > 0$ sufficiently small we can make $s_i + \delta h \cdot \nabla f^i(x^*) \geq 0$. Therefore,

$$s_i + b_i + \delta h \cdot \nabla f^i(x^*) \geq b_i.$$

Finally, $f^0(x^* + \delta h) > f^0(x^*)$ using the Taylor series and the fact that $-h \cdot \nabla f^0(x^*) < 0$, which contradicts local optimality of x^*. ■

The LI constraint qualification is sensitive to the representation of the constraint set. To see why, consider the problem $\max\{f^0(x): g(x) = 0\}$. This can be rewritten as $\max\{f^0(x): g(x) \geq 0, -g(x) \geq 0\}$. Notice that the set of vectors $\{\nabla g(x), -\nabla g(x)\}$ is not LI so Theorem 5.4 does not apply (at least not directly) to problems with equality constraints.

Nevertheless, an extension of the Kuhn–Tucker–Karush theorem to equality constraints is available. Append to the index set M of constraints in problem (P) an additional set $M^= = \{i \in M: f^i(x) = 0\}$. Let (P') be the following problem:

$$\max\{f^0(x): f^i(x) \geq 0 \; \forall i \in M, \; f^i(x) = 0 \; \forall i \in M^=\}. \tag{P'}$$

Theorem 5.6 *Let x^* be a local maximum for problem (P') (which now has some equality constraints). Suppose the functions f_i for all $i \in \{0\} \cup \{M \cup M^=\}$ and their first derivatives are continuous. If the vectors in $\{\nabla f^i(x^*)\}$ for $i \in \{M^= \cup \{i \in M: f^i(x^*) = 0\}\}$ are LI there exist multipliers $\{\lambda_i\}_{i \in M \cup M^=}$ such that*

1. $\nabla f^0(x^*) + \sum_{i \in M \cup M^=} \lambda_i \nabla f^i(x^*) = 0$,
2. $\lambda_i \geq 0 \; \forall i \in M$,
3. λ_i *unrestricted for $i \in M^=$*,
4. $\lambda_i f^i(x^*) = 0 \; \forall i \in M$,
5. $f^i(x^*) \geq 0 \; \forall i \in M$ *and*
6. $f^i(x^*) = 0 \; \forall i \in M^=$.

Proof We will prove the existence of multipliers μ_i for $i \in \{0\} \cup \{M \cup M^=\}$ not all zero such that

$$\mu_0^* \nabla f^0(x^*) + \sum_{i \in M \cup M^=} \mu_i^* \nabla f^i(x^*) = 0. \tag{5.3}$$

The remainder of the proof will follow the proof of Theorem 5.4.

Let $F' = \{x \in \mathbb{R}^n: f^i(x) \geq 0 \; \forall i \in M \cup M^=\}$, $g(x) = \sum_{i \in M^=} f^i(x)$ and and $G = \{x \in \mathbb{R}^n: g(x) \leq 0\}$. Problem (P') can be reformulated as $\max\{f^0(x): x \in F' \cap G\}$.

Since x^* is a local maximum for problem (P') there is an $\epsilon > 0$ such that x^* solves

$$\max\{f^0(x): x \in N_\epsilon(x^*) \cap G\}.$$

Let $f(x) = f^0(x) - f^0(x^*) - \|x - x^*\|^2$. Observe that $f(x) \leq 0$ for all $x \in N_\epsilon(x^*) \cap G$ with equality iff $x = x^*$. Since the gradients of f^0 and f coincide at x^* it suffices to to prove (5.3) with f^0 replaced by f.

Choose $\delta < \epsilon$ and let $D = \{x: d(x, x^*) < \delta\}$, $D' = \{x: d(x, x^*) \leq \delta\}$ and $B = \{x: d(x, x^*) = \delta\}$ where $\delta < \epsilon$. Subsequently δ will be sent to zero. We partition $F' \cap D'$ into three sets: $F^1 = F' \cap D$, $F^2 = \{x \in F' \cap B: f(x) < 0\}$ and $F^3 = \{x \in F' \cap B: f(x) \geq 0\}$. Since x^* is not a local maximum for $\max\{f(x): x \in F'\}$, $F^3 \neq \emptyset$.

Since $x^* \notin B$ and $B \subset N_\epsilon(x^*)$ it follows that $g(x) > 0$ for all $x \in F^3$. Since $f(x)/g(x)$ is well defined for all $x \in F^3$ and F^3 is compact, there is a positive number $\theta > f(x)/g(x)$ for all $x \in F^3$. If F^3 is empty, choose any positive θ.

Next, $f(x) < 0 \leq g(x)$ for all $x \in F^2$. In addition, $f(x^*) = 0 = g(x^*)$ and $x^* \in F^1$.

Consider the function $f(x) - \theta g(x)$. It is strictly negative for all $x \in F^2 \cup F^3$ and non-negative for at least one $x \in F^1$. By the Weierstrass theorem there is a $z \in F' \cap D'$ that maximizes $f(x) - \theta g(x)$. Since $z \in F^1$, $\|z - x^*\| < \delta$. Also, z is a local maximum for $\max\{f(x) - \theta g(x): x \in F'\}$. From Lemma 5.3 we deduce the existence of non-negative multipliers $v_0, v_1, v_2, \ldots$ not all zero such that

$$v_0 \nabla[f(z) - \theta g(z)] + \sum_{i \in M \cup M^+} v_i \nabla f^i(z) = 0. \tag{5.4}$$

We can rewrite (5.4) as

$$\mu_0 \nabla f(z) + \sum_{i \in M \cup M^+} \nabla \mu_i f^i(z) = 0, \tag{5.5}$$

where $\mu_i = v_i$ for all $i \in M \cup \{0\}$ and $\mu_i = v_i - \theta v_0$ for all $i \in M^=$. Since the v's are not all zero, the μ's are not all zero. By scaling we can assume that $\sum_{i \in \{0\} \cup M \cup M^=} |\mu_i| = 1$.

Now let δ tend to zero. Since $z \in N_\delta(x^*)$ it follows that $z \to x^*$. Since the μ's are bounded, by the Bolzano–Weierstrass theorem, there is a convergent subsequence of them tending to a limit $\mu_0^*, \mu_1^*, \mu_2^*, \ldots$. Finally, continuity of the gradients of our functions imply that equation (5.5) holds as $z \to x^*$.

Item (2) follows from the fact that $\mu_i = v_i \geq 0$ for all $i \in M$. Note that we can say nothing about the sign of μ_i for $i \in M^=$ since $\mu_i = v_i - \theta v_0$ and this could be negative. Item (4) is complementary slackness. To derive it let $M^+ = \{i \in M: f^i(x^*) > 0\}$. Then x^* is a local maximum for the problem

$$\max\{f^0(x): f^i(x) \geq 0 \ \forall i \in M \setminus M^+, f^i(x) = 0 \ \forall i \in M^=\}.$$

In words, problem P' with the constraints in M^+ omitted. Now repeat the argument for this problem. Since the constraints associated with M^+ do not appear, this is equivalent to setting their multipliers to zero. Items (5) and (6) follow from feasibility of x^*. ∎

Condition 1 in Theorem 5.6 is usually called a **first-order conditions** and the multipliers associated with $i \in M^=$ are called **Lagrange multiplier's**. Theorem 5.6 describes equations that a local maximum must satisfy. It does not follow that every solution of this system of equations is a local maximum.

Example 20 *Consider the following optimization problem:*

$$\max x_1^2 + x_2^2$$
$$\text{s.t.} \quad x_1 + x_2 - 1 \geq 0.$$

Here $f^0 = x_1^2 + x_2^2$ and $f^1 = x_1 + x_2 - 1$. The equation

$$\nabla f^0(x^*) + \sum_{i \in M \cup M^=} \lambda_i \nabla f^i(x^*) = 0,$$

and complementary slackness gives rise to

$$2x_1 + \lambda = 0,$$
$$2x_2 + \lambda = 0,$$
$$\lambda(x_1 + x_2 - 1) = 0.$$

The first two equations yield $x_1 = -\lambda/2 = x_2$. Substituting this into the complementary slackness condition gives $\lambda(-\lambda - 1) = 0$. There are two solutions: $\lambda = 0$ and $\lambda = -1$. We can discard the $\lambda = -1$ solution since λ must be non-negative; recall we have an inequality constraint. This leaves $x_1 = x_2 = 0$ as the only solution, which is not a local maximum.

Example 21 *Consider the following optimization problem:*

$$\max \ -2x_1^2 - 3x_2^2$$
$$\text{s.t.} \quad x_1 + x_2 - 1 = 0.$$

Here $f^0 = -2x_1^2 - 3x_2^2$ and $f^1 = x_1 + x_2 - 1$. The equation

$$\nabla f^0(x^*) + \sum_{i \in M \cup M^=} \lambda_i \nabla f^i(x^*) = 0,$$

and feasibility give rise to

$$-4x_1 + \lambda = 0,$$
$$-6x_2 + \lambda = 0,$$
$$x_1 + x_2 = 1.$$

The system has a unique solution; $\lambda = 12/5$ and $x_1 = 3/5$ and $x_2 = 2/5$. If our given optimization problem has a solution, it must be the one we have identified above. Notice the qualifier 'if'. One must prove that the problem at hand has an optimal solution before concluding that $x_1 = 3/5$ and $x_2 = 2/5$ is its optimal solution.

5.2 Sufficient conditions for optimality

In order for the necessary conditions identified above to be sufficient for global optimality we must make assumptions about the shape of the objective function f^0.

5.2.1 Concave functions

Definition 5.7 *Let C be a convex subset of $\mathbb{R}^n$ and $f: C \to \mathbb{R}$. The function f is said to be **concave** if $f(\lambda x + (1 - \lambda)y) \geq \lambda f(x) + (1 - \lambda)f(y)$ for all $x, y \in C$ and $\lambda \in [0, 1]$.*

The function f is **convex** if $-f$ is concave. In the sequel we concentrate on concavity and leave the reader to make the appropriate changes for the convex

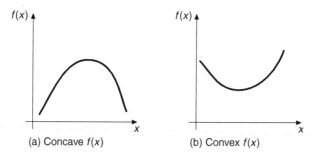

(a) Concave f(x)   (b) Convex f(x)

Figure 5.3

case. Graphical illustrations of a concave and convex function in one real variable are shown in Figure 5.3. Concavity is a very strong property as can be seen from the next result.

Theorem 5.8 *Let $C \subset \mathbb{R}^n$ be an open convex set and $f: C \to \mathbb{R}$ a concave function such that $\max_{x \in C}|f(x)| \leq M$ for some constant M. Then f is continuous on C.*

Proof Set $g(x) = -f(x)$. Then g is convex on C. We will show that g is continuous. Pick any $z \in C$. By a suitable translation we can assume that $z = 0$ and $g(z) = 0$. Simply consider $x - z$ and the function $g(x) - g(z)$.

Consider the point 0 and the ball B of radius r around it that is contained in C. Such a ball exists because C is open. Now choose an ϵ such that $1 > r > \epsilon > 0$, and consider any x whose distance from 0 is at most $\epsilon \times r$. This ensures that $x/\epsilon \in B$ (see Figure 5.4). For any such x we have $g(x) = g[(1 - \epsilon) \times 0 + \epsilon \times (x/\epsilon)] \leq \epsilon g(x/\epsilon)$ by convexity. Since $g(x) \leq M$ for all $x \in C$ we deduce that $g(x) \leq \epsilon M$. Next,

$$0 = g(0) = g\left[(1 + \epsilon)^{-1}x + \epsilon(1 + \epsilon)^{-1}\left(-\frac{x}{\epsilon}\right)\right]$$
$$\leq (1 + \epsilon)^{-1}g(x) + \epsilon(1 + \epsilon)^{-1}M.$$

So, $g(x) \geq -\epsilon M$. Hence for all $x \in C$ such that $|x - 0| \leq r < \epsilon$, we have $|g(x) - g(0)| \leq \epsilon M$. ∎

Concave functions have many useful properties and equivalent definitions. These are listed below and are easy to prove:

1. Let $C \subset \mathbb{R}^n$ be convex. $f: C \to \mathbb{R}$ is concave iff $\{(x, t) \in \mathbb{R}^{n+1}: x \in S, t \leq f(x)\}$ is convex. The set $\{(x, t) \in \mathbb{R}^{n+1}: x \in S, t \leq f(x)\}$ is called the **hypograph** of f.
2. If f is a concave function on a convex set C then $h(x) = \mu f(x)$ is a concave function on C for all $\mu \geq 0$.
3. If f and g are concave functions on a convex set C then $h(x) = f(x) + g(x)$ is a concave function on C.

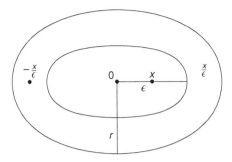

Figure 5.4

4. If f and g are concave functions on a convex set C then $h(x) = \min\{f(x), g(x)\}$ is a concave function on C.
5. If f is a concave function on a convex set C. Let $I = \{t : \exists x \in C \text{ s.t. } t = f(x)\}$ and u be a non-decreasing concave function on I. Then $h(x) = u[f(x)]$ is concave on C.
6. If f is a concave function on a convex set C and $\{x^1, \dots, x^n\}$ are points in C then

$$f\left(\sum_{i=1}^{n} \lambda_i x^i\right) \geq \sum_{i=1}^{n} \lambda_i f(x^i)$$

for all $\{\lambda_i\}_{i \geq 1} \geq 0$ such that $\sum_{i=1}^{n} \lambda_i = 1$.

Imposing differentiability on a concave function allows one to characterize them in simple ways. One example follows.

Definition 5.9 *Let $f : \mathbb{R}^n \to \mathbb{R}$ be a continuous twice differentiable function (meaning all its second derivatives exist). The **Hessian** of f at x called $H_f(x)$ is the $n \times n$ matrix whose $\{ij\}$th entry is $\frac{\partial^2 f}{\partial x_i \partial x_j}$.*

Theorem 5.10 *Let $C \subset \mathbb{R}^n$ be convex and f a continuous, twice differentiable function on C. Then f is concave iff $u^T H_f(x) u \leq 0$ for all $x \in C$ and $u \in \mathbb{R}^n$.*

An $n \times n$ matrix A is said to be **negative semi-definite** if $u^T A u \leq 0$ for all $u \in \mathbb{R}^n$. Checking that a matrix is semi-definite is not easy, but there are ways to do so using the signs of the determinants of various square submatrices of A.[1]

Concave functions that map an interval I of the real line into $\mathbb{R}^1$ are also quite useful.

Lemma 5.11   *Let I be an interval of* $\mathbb{R}^1$ *and* $f: I \to \mathbb{R}$. *Then f is concave iff*

$$\frac{f(x_2) - f(x_1)}{x_2 - x_1} \geq \frac{f(x_3) - f(x_2)}{x_3 - x_2}$$

for any x_1, x_2 *and* $x_3 \in I$ *such that* $x_1 < x_2 < x_3$.

Remark   To understand the lemma it is useful to refer to Figure 5.5. The lemma asserts that the slope of the line segment joining $(x_1, f(x_1))$ and $(x_2, f(x_2))$ exceeds the slope of the line segment joining $(x_2, f(x_2))$ and $(x_3, f(x_3))$.

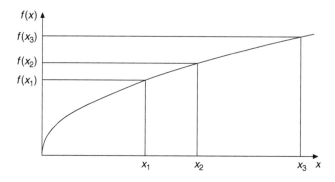

Figure 5.5

Proof   Assume the inequality in the statement of the theorem holds. Fix an $x_1 \in I$ and $x_3 \in I$ with $x_1 < x_3$. For any $0 < \lambda < 1$ we can choose x_2 so that $x_2 = \lambda x_1 + (1 - \lambda)x_3$. Since $x_1 < x_2 < x_3$:

$$(x_3 - x_1)f(x_2) \geq (x_3 - x_2)f(x_1) + (x_2 - x_1)f(x_3).$$

Since $\lambda = (x_3 - x_2)/(x_3 - x_1)$ we can rewrite the previous inequality as

$$f(x_2) \geq \lambda f(x_1) + (1 - \lambda)f(x_3).$$

Thus f is concave.

Now suppose f is concave. Given $x_1 < x_2 < x_3$ choose λ so that $x_2 = \lambda x_1 + (1 - \lambda)x_3$. And work the previous argument in reverse.   ∎

Lemma 5.12   *Let I be an interval of* $\mathbb{R}$ *and* $f: I \to \mathbb{R}$ *be differentiable. Then f is concave iff* $f'(u) \geq f'(v)$ *for any* $u, v \in I$ *such that* $u < v$.

Proof Suppose f concave and choose $\delta < (v - u)/2$. Applying Lemma 5.11

$$\frac{f(u + \delta) - f(u)}{\delta} \geq \frac{f(v - \delta) - f(u + \delta)}{v - u - 2\delta} \geq \frac{f(v) - f(v - \delta)}{\delta}.$$

Let $\delta \to 0$, and we deduce that $f'(u) \geq f'(v)$. Now suppose that f is not concave. Then we can choose three numbers $x_1 < x_2 < x_3$ in I to violate Lemma 5.11. By Rolle's theorem we can choose u and v so that

1. $x_1 < u < x_2 < v < x_3$,
2. $(x_2 - x_1) f'(u) = f(x_2) - f(x_1)$,
3. $(x_3 - x_2) f'(v) = f(x_3) - f(x_2)$.

Hence $f'(u) < f'(v)$ a contradiction. ∎

Lemma 5.12 implies:

Theorem 5.13 *Let I be an interval of $\mathbb{R}$ and $f: I \to \mathbb{R}$ be twice differentiable. Then f is concave iff $f''(x) \leq 0$ for all $x \in I$.*

5.2.2 Concave programming

The concave programming problem is problem (P) when f^0 and $\{f^i\}_{i \in M}$ are all concave. Notice that in this case the feasible region is convex. We denote the concave programming problem by P_c. We consider the unconstrained case first.

Theorem 5.14 (Unconstrained case) *Let f be a concave, continuous, differentiable function on an open convex set C. Then f has a maximum at $x^* \in C$ iff $\nabla f(x^*) = 0$.*

Proof Since a global maximum is a local maximum, one direction follows from Theorem 5.2. To prove the other direction let $x \neq y$ be such that $f(x) > f(y)$. We show that $\nabla f(y) \neq 0$. Concavity of f implies

$$f(\mu x + (1 - \mu)y) \geq \mu f(x) + (1 - \mu) f(y),$$

whenever $0 < \mu < 1$. Set $h = x - y$ and $\theta = f(x) - f(y) > 0$. Then the inequality above can be rewritten as

$$f(y + \mu h) - f(y) \geq \mu \theta.$$

From the Taylor series expansion of f it follows that for all sufficiently small $\mu > 0$, $h \cdot \nabla f(y) \geq \theta > 0$. Hence $\nabla f(y) \neq 0$. ∎

Definition 5.15 *A point x^* is called a **Kuhn–Tucker–Karush** point for problem (P_c) if there exist multipliers $\{\lambda_i\}_{i \in M}$ such that:*

1. $\nabla f^0(x^*) + \sum_{i \in M} \lambda_i \nabla f^i(x^*) = 0$
2. $\lambda_i f^i(x^*) = 0, \forall i \in M$

3. $\lambda_i \geq 0, \forall i \in M$
4. $f^i(x^*) \geq 0, \forall i \in M$.

Theorem 5.16 (Constrained case) *If x^* is a Kuhn–Tucker–Karush (KTK) point for (P_c), then x^* is an optimal solution to (P_c).*

Proof Observe first that $f^0(x) + \sum_{i \in M} \lambda_i f^i(x)$ is a non-negative combination of concave functions and so is concave. By the first condition of being a KTK point and Theorem 5.14, x^* maximizes $f^0(x) + \sum_{i \in M} \lambda_i f^i(x)$. The last condition of being a KTK point implies that x^* is feasible for problem (P_c). Now pick any other feasible solution, x to (P_c). Then

$$f^0(x^*) + \sum_{i \in M} \lambda_i f^i(x^*) \geq f^0(x) + \sum_{i \in M} \lambda_i f^i(x).$$

By the second condition of being a KTK point, $\sum_{i \in M} \lambda_i f^i(x^*) = 0$. Hence

$$f^0(x^*) \geq f^0(x) + \sum_{i \in M} \lambda_i f^i(x) \geq f^0(x).$$

The last inequality follows from the third condition of being a KTK point. ∎

The reader will wonder why we have not considered the case of equality constraints. If a constraint function f is concave, the set $\{x: f(x) \geq 0\}$ is convex. If we have an equality constraint $f(x) = 0$ we can replace it by the inequalities $f(x) \leq 0$ and $f(x) \geq 0$. However, the set $\{x: f(x) \leq 0\}$ is not convex. So, the region determined by $\{x: f(x) \geq 0\} \cap \{f(x) \leq 0\}$ need not be convex. Notice also, that this difficulty vanishes when f is linear.

5.2.3 Constraint qualifications

Here we summarize the three most popular constraint qualifications:

1. In problem (P'), the gradients $\{\nabla f^i(x^*)\}_{i \in M \cup M^=}$ are linearly independent.
2. In problem (P') the constraint functions $\{f^i\}_{i \in M \cup M^+}$ are linear.
3. In problem (P_c) there exists a feasible x such that $f^i(x) > 0$ for all $i \in M$.

Only the first depends on x^*. The third one, called the **Slater condition**, does not apply when equality constraints are present.

5.3 Envelope theorem

Frequently one is interested in the change in optimal objective function value as one changes some parameter. This parameter can be in the objective function, the constraints or both. We have already seen an example of this with the marginal value theorem. For non-linear optimization problems the tool of choice is the envelope theorem. This theorem should really be viewed as the generalization of the marginal value theorem to non-linear optimization problems.

Let $f: \mathbb{R}^n \times [0, 1] \to \mathbb{R}$ be a parameterized objective function. Let F be a feasible set of points and

$$V(t) = \max\{f(x, t): x \in F\}.$$

Notice that the definition of $V(t)$ assumes that f attains a maximum in F. What we would like to know is the derivative of $V(t)$ with respect to t. If we fix a value of t, we can rewrite $V(t)$ as follows:

$$V(t) = \min\{y: y \geq f(x, t), \ \forall x \in F\}.$$

Thus $V(t)$ is the optimal objective function value of a linear program, albeit one with as many variables as there are points in F. Changing the value of t amounts to a change in the right hand side of this last program. Thus, from the marginal value theorem, we would conjecture that $V(t + \epsilon) - V(t)$ should depend on how $f(x, t)$ changes with t.

To get a sense of what kind of theorems one can expect, suppose that F consists of a finite number of points and that f is differentiable in t. Recall that

$$V(t) = \min \ y$$
$$\text{s.t.} \quad y \geq f(x, t), \quad \forall x \in F.$$

This is a linear program. If f has a unique maximizer, x^*, in F, there will be exactly one binding constraint in the optimal solution. Now suppose we would like, from the solution to this linear program, the value of $V(t + \epsilon)$.

We can approximate $f(x, t + \epsilon)$ in the neighborhood of t by its Taylor series expansion. So we can replace $f(x, t + \epsilon)$ by $f(x, t) + \epsilon f_t(x, t)$. Here f_t denotes the derivative of f with respect to t. Hence

$$V(t + \epsilon) = \min \ y$$
$$\text{s.t.} \quad y \geq f(x, t) + \epsilon f_t(x, t), \quad \forall x \in F.$$

Thus computing $V(t + \epsilon)$ amounts to increasing the right hand side of each constraint in the original program by $\epsilon f_t(x, t)$. If ϵ is small enough, the constraint associated with x^* continues to be the only one to bind. So

$$V(t + \epsilon) = f(x^*, t) + \epsilon f_t(x^*, t).$$

Hence

$$\frac{V(t + \epsilon) - V(t)}{\epsilon} = f_t(x^*, t).$$

Letting $\epsilon \to 0$ gives $V_t(t) = f_t(x^*, t)$. This is an example of an envelope theorem.

To see how all of this is consistent with the marginal value theorem, return to the original program:

$$V(t) = \min \ y$$
$$\text{s.t.} \ \ y \geq f(x, t), \quad \forall x \in F.$$

The dual is

$$\max \sum_{x \in F} f(x, t)\mu_x$$

$$\text{s.t.} \ \sum_{x \in F} \mu_x \leq 1,$$

$$\mu_x \geq 0, \quad \forall x \in F.$$

It is easy to see that the optimal solution to the dual is $\mu_{x^*} = 1$ and $\mu_x = 0$ for all $x \neq x^*$ and this is unique. When we increase t by ϵ, the right hand sides of the original program change by $\epsilon f_t(x, t)$, so, by the marginal value theorem

$$V(t + \epsilon) = V(t) + \sum_{x \in F} \mu_x f_t(t, x) = V(t) + f_t(x^*, t).$$

The whole trick in proving envelope theorems is to invoke conditions that ensure that the constraint set $\{y: y \geq f(x, t) \ \forall x \in F\}$ is well behaved for small changes in t. Specifically, a constraint that was binding at the optimal solution continues to be binding when we increase t to $t + \epsilon$ for ϵ sufficiently small.

We close this section with one instance of the envelope theorem.[2] Let $t \in \mathbb{R}^k$ and

$$V(t) = \max \ f^0(x, t)$$
$$\text{s.t.} \ \ f^i(x, t) = 0, \quad \forall i \in M.$$

All functions are concave. For each t, let $x(t)$ be an optimal solution and suppose the constraint qualification holds. Associated with each $x(t)$ are a set of multipliers $\{\lambda_i(t)\}_{i \in M}$ such that

$$\nabla f^0(x(t), t) + \sum_{i \in M} \lambda_i(t) \nabla f^i(x(t), t) = 0.$$

Theorem 5.17   *Suppose $x(t)$ is a differentiable function of t. Then*

$$\nabla V(t) = \nabla_t f^0(x(t), t) - \sum_{i \in M} \lambda_i(t) \nabla_t f^i(x(t), t).$$

Proof   By the chain rule

$$\frac{\partial V(t)}{\partial t_s} = \frac{\partial f^0(x(t),t)}{\partial t_s} + \sum_{j=1}^{n} \left(\frac{\partial f^0(x(t),t)}{\partial x_j} \frac{\partial x(t)}{\partial t_s} \right).$$

From the first order condition we have:

$$\frac{\partial f^0(x(t),t)}{\partial x_j} = -\sum_{i \in M} \lambda_i(t) \frac{\partial f^i(x(t),t)}{\partial x_j}.$$

Hence

$$\frac{\partial V(t)}{\partial t_s} = \frac{\partial f^0(x(t),t)}{\partial t_s} - \sum_{i \in M} \lambda_i(t) \sum_{j=1}^{n} \left(\frac{\partial f^i(x(t),t)}{\partial x_j} \frac{\partial x(t)}{\partial t_s} \right).$$

However, $f^i(x(t),t) = 0$ for all t. Thus

$$\sum_{j=1}^{n} \frac{\partial f^i(x(t),t)}{\partial x_j} \frac{\partial x(t)}{\partial t_s} = \frac{\partial f^i(x(t),t)}{\partial t_s}$$

and this proves the theorem.   ∎

5.4   An aside on utility functions

Economics starts with the assumption that agents are rational; suggesting that mad dogs and Englishmen are not the only ones who go out into the noonday sun. Its an assumption that attracts criticism the way horse shit attracts flies and deserves discussion, but not here.

There are two parts to the definition of rationality used in Economics. The first is that agents are defined by their **preferences** over things or outcomes. Further, these preferences satisfy consistency conditions described below.

First, for any *any* two bundles of goods and services, call them x and y, our agent should be able to say exactly one of the following:

1.   She prefers x to y.
2.   She prefers y to x.
3.   She is indifferent between x and y.

If she can do this, we say she has a **preference ordering** over the set of all bundles. We don't care how she arrives at this ordering, only that she has one. This is how economics differs from, say, sociology. Preferences are assumed to be fixed and innate. Why someone's preferences are the way they are is not, for our purposes, relevant.[3] Preferences are required to satisfy three conditions:

1.   **Monotonicity:** More of a good thing is better (and certainly no worse) than less of it.
2.   **Irreflexivity:** Given two identical bundles, you should never prefer one to the other.
3.   **Transitivity:** If x is preferred to y and y is preferred to z *then* x is preferred to z. If I prefer apples to oranges and oranges to grapefruit, then I prefer apples to grapefruit.

One other requirement, invoked for convenience, is called the **law of diminishing returns**.[4] The benefit derived from successive units of a particular commodity diminish as total consumption of that commodity increases, the consumption of all other commodities being held constant. The more salt you have, the less additional salt you want.

An agent will be **consistent** if their preference orderings conform to the above. The second part of the definition stipulates how a consistent agent chooses between bundles of goods and services.

Given a set of bundles to choose from, the consistent agent will choose their most preferred bundle from the set. Writing in 1881, Francis Ysidro Edgeworth (1845–1926), put it thus: 'the first principle of Economics is that every agent is actuated only by self-interest'.[5] The rational agent looks only at their preferences and no one else's in deciding on the best bundle. This narrow view of human behavior is mistakenly ascribed to Adam Smith (1723–1790)[6] as the following limerick by Stephen Leacock[7] suggests.

> *Adam, Adam, Adam Smith*
> *Listen what I charge you with!*
> *Didn't you say*
> *In the class one day*
> *That selfishness was bound to pay?*
> *Of all doctrines that was the Pith,*
> *Wasn't it, wasn't it, wasn't it, Smith?*

A preference ordering is awkward to write down. It would be useful to have a compact representation of it. A numerical representation of a preference ordering over the set of bundles is a function u such that

$$x \text{ is preferred to } y \text{ if and only if } u(x) \geq u(y) \tag{5.6}$$

for all x and y. The function u is called a **utility function**. Basically, one can assign a numerical score to each bundle with the property that more preferred bundles get a higher score. The score that is assigned to a particular bundle represents the utility to be had from that bundle.

If the preference ordering satisfies the four conditions listed above, then it can be represented by a *non-decreasing* and *concave* utility function. Given a utility function, the rational agent chooses the bundle with highest or maximum utility.

It is important to remember that a utility function does nothing more than represent preferences. It tells us nothing about the intensity of preferences. The fact that $u(x) = 7$ and $u(y) = 3$ tells us nothing about much more an agent with this utility function prefers x to y. To see why this is the case, observe that if $u(\cdot)$ is a utility function representing some preference order than $\lambda u(\cdot)$ where $\lambda > 0$ represents the same of preferences.

The utility framework can be extended to choice in an uncertain world by extending the notion of bundles of commodities to include 'lotteries'. The word is interpreted in the broad sense to include any risky choice. For example, a hundred shares in IBM to be sold two weeks from now is like a lottery ticket. The profit is uncertain and beyond your control.[8] A prospective employee is a lottery ticket. She may turn out to be wonderful or a real dolt. You may be able to guess which with some confidence, but you don't know for certain. We can represent all such risky prospects as lottery tickets which payoff particular amounts with a particular probability. We require given any two lottery tickets, that one specify which you prefer over the other or whether you are indifferent. By imposing consistency conditions on the ordering of lotteries we can derive a utility function representation. Furthermore, this utility function has the property that the utility of a lottery is just the expected utility of its different outcomes.

One useful number that we can associate with a lottery is it's expected payoff. The expected payoff is a useful benchmark for classifying an individuals attitude toward risk. Here is how.

Assume you own a lottery ticket which will pay $7 with a probability $1/2$ and zero otherwise. Suppose someone offers to buy it from you. If you are willing to sell it for $3.50 or less, you are **risk averse**. If you will only sell it for something more than $3.50, you are **risk seeking**. When you think it's worth exactly $3.50 no more and no less, you are **risk neutral**.

An agent who is risk averse is modeled using a concave utility function. To see why, suppose a lottery ticket that pays x with probability λ and pays y with probability $1 - \lambda$. The expected payoff of the ticket is $\lambda x + (1 - \lambda)y$. Suppose an agent with a concave utility function, who is offered a choice between the lottery ticket and a sure payoff of $\lambda x + (1 - \lambda)y$. The expected utility of the lottery ticket to the agent is $\lambda u(x) + (1 - \lambda)u(y)$ which by concavity is at most $u(\lambda x + (1 - \lambda)y)$. Thus the utility of the sure thing is at least as large as the utility of the lottery. In words the agent would prefer the sure thing to the lottery.

5.5   Application: market games

We consider an economy with k divisible goods and a set N of agents. Each agent is endowed with a non-negative quantity of each good that we represent as a vector in $\mathbb{R}^k_+$. The endowment of agent i is $w^i \in \mathbb{R}^k_+$. Each agent i is also endowed with a non-negative amount $\hat{m}_i$ of money. Agent i's preferences over a vector of goods are represented by a continuous, concave utility function $u^i : \mathbb{R}^k \rightarrow \mathbb{R}$. We assume that utilities for all agents are **quasi-linear**. Thus the utility assigned by agent i to a bundle $x \in \mathbb{R}^k_+$ of goods and an amount m of money is $u^i(x) + m$. The assumption of quasi-linearity means that the utilities of all agents can be measured on a common monetary scale. An implication of this is that utility can be transferred from one agent to another through the medium of money.

The only transactions permitted in this economy are exchanges or trades of goods. The transactions we expect to see are those that make every participant in the transaction at least as well as off as before transacting. For example, if one agent has apples only but prefers oranges, while the other has oranges but prefers apples, they would both be better of if they were to swap some apples for oranges. Even if there are gains from trade it does not follow that those gains will be realized. The agents must still haggle over how those gains are to be divided amongst themselves. Trade could break down if the agents reach no agreement on the division.

We will assume that when a group S of agents meet to trade amongst themselves only, they will trade in such a way as to maximize the sum of their utilities. Implicit is the assumption that the agents will reach an agreement on how the gains are to be divided. The question we answer is this: what will the resulting distribution of utilities look like.

Given a subset S of agents we formulate the problem of redistributing their initial endowment of goods and money so as to maximize their total utility as a concave programming problem.

Let x^i be the vector of goods assigned to agent $i \in S$ and m^i the change in monetary position. If $m^i > 0$, agent i receives money, if $m^i < 0$ then agent i pays out and if $m^i = 0$ agent i's monetary position is unchanged. For trades to be feasible the following constraint must hold:

$$\sum_{i \in S} x^i = \sum_{i \in S} w^i,$$

$$\sum_{i \in S} m^i = 0.$$

The last constraint follows from the fact that sum total of money exchanged must be zero.

The maximum total utility that the players in S can achieve is $v(S)$ and

$$v(S) = \max \sum_{i \in S} u^i(x^i) + \sum_{i \in S} (\hat{m}^i + m^i)$$

$$\text{s.t.} \sum_{i \in S} x^i = \sum_{i \in S} w^i,$$

$$\sum_{i \in S} m^i = 0,$$

$$x^i \in \mathbb{R}^k_+.$$

Because of the last constraint we can ignore the terms involving money and just set

$$v(S) = \max \left\{ \sum_{i \in S} u^i(x^i): \sum_{i \in S} x^i = \sum_{i \in S} w^i \right\}.$$

The cooperative game defined by this value function v is called a *market* game. Notice that $v(S) \leq v(T)$ whenever $S \subset T$. Thus the gains from trade (as measured by total utility) increase with the number of agents involved. The largest possible gains occurs when all agents in N trade amongst themselves. The core of this game, if it exists, is a reasonable of prediction of the set of possible utility distributions. If the result of trading was a distribution of utilities that lay outside the core, there would be a subset of agents who could get together and do better.

Theorem 5.18   *If v is a market game then $C(v, N) \neq \varnothing$.*

Proof   For each $S \subset N$ let x_S^i satisfy $v(S) = \sum_{i \in S} u^i(x_S^i)$ and $\sum_{i \in S} x_S^i = \sum_{i \in S} w^i$. We know that such x_S^i's exist because we are maximizing a continuous function over a compact set, so the set must contain an optimum.

Pick a $y \in B(N)$. Let $z^i = \sum_{S \ni i} y_S x_S^i$. We show that $\{z^i\}_{i \in N}$ is a feasible allocation for all N players, i.e., $\sum_{i \in N} z^i = \sum_{i \in N} w^i$. Now

$$\sum_{i \in N} z^i = \sum_{i \in N} \sum_{S \ni i} y_S x_S^i = \sum_{S \subset N} y_S \sum_{i \in S} x_S^i =$$

$$= \sum_{S \subset N} y_S \sum_{i \in S} w^i = \sum_{i \in N} w^i \sum_{S \ni i} y_S = \sum_{i \in N} w^i$$

since $\sum_{S \ni i} y_S = 1$.

Now that we have a feasible solution z for the entire economy we can use the Bondareva–Shapley theorem to show that the core is non-empty.

We know that

$$v(N) = \max \left\{ \sum_{i \in N} u^i(t^i): \sum_{i \in N} t^i = \sum_{i \in N} w^i \right\} \geq \sum_{i \in N} u^i(z^i).$$

By concavity of the utility functions

$$v(N) \geq \sum_{i \in N} u^i \left(\sum_{S \ni i} y_S x_S^i \right) \geq \sum_{i \in N} \sum_{S \ni i} y_S u^i(x_S^i)$$

$$= \sum_{S \subset N} y_S \sum_{i \in S} u^i(x_S^i) = \sum_{S \subset N} y_S v(S),$$

since $v(S) = \sum_{i \in S} u^i(x_S^i)$. The theorem now follows from the Bonderava–Shapley theorem. ∎

Trades are typically conducted with prices. It is natural to ask if there is a set of prices that would lead to a reallocation of the endowments that is in the core.

Let $p \in \mathbb{R}^k$ be a price vector. The price vector is unrestricted in sign. If the price of a good is negative, it means that someone must be paid to buy it. Given p, agent i solves the following optimization problem to determine what to ask for.

$$\max \quad u^i(x^i) + \hat{m}^i + m^i$$
$$\text{s.t.} \quad px^i + \hat{m}^i + m^i = \hat{m}^i + p\omega^i,$$
$$x^i \geq 0.$$

Dropping constant terms the agents optimization problem can be simplified to

$$\max \quad u^i(x^i) + m^i$$
$$\text{s.t.} \quad px^i + m^i = p\omega^i,$$
$$x^i \geq 0.$$

The feasible region is compact and the objective function continuous, so by the Weierstrass theorem an optimal solution exists. Substituting the one constraint into objective functions yields:

$$\max \quad u^i(x^i) + p(\omega^i - x^i)$$
$$\text{s.t.} \quad x^i \geq 0.$$

Since $p\omega$ is constant we can drop it from the optimization problem and reduce it to

$$\max \quad u^i(x^i) - px^i \quad \text{s.t.} \quad x^i \geq 0.$$

Denote the optimal solution by $x^i(p)$. Note the dependence on p.

A price vector p is an **equilibrium** for the market if demand equals supply, i.e.

$$\sum_{i=1}^{n} x^i(p) = \sum_{i=1}^{n} \omega^i.$$

An optimal solution to each agents optimization problem must satisfy the KTK condition

$$\frac{\partial u^i}{\partial x^i_j} - p_j + \mu^i_j = 0$$

for each good j. Here μ^i_j is the multiplier associated with the constraint $x^i_j \geq 0$.

If $x_j^i(p) > 0$ then $\mu_j^i = 0$ by complementary slackness. In this case

$$\frac{\partial u^i}{\partial x_j^i} - p_j = 0.$$

If $x_j^i(p) = 0$, we know only that $\mu_j^i \geq 0$. Hence

$$\frac{\partial u^i}{\partial x_i} - p_i \leq 0.$$

Now imagine a benevolent planner that tries to allocate the resources of this economy so as to maximize the sum of utilities, i.e., the planner computes $v(N)$. The planners problem is to choose $\{z^i\}_{i \in N}$ so that

$$\sum_{i \in N} u^i(z^i) = \max \left\{ \sum_{i \in N} u^i(x^i) : \sum_{i \in N} x^i = \sum_{i \in N} \omega^i, x^i \geq 0 \right\}.$$

Since this is a concave programming problem, it follows from the KTK conditions that

$$\frac{\partial u^i}{\partial x_j^i} \bigg|_{x_j^i = z_j^i} - \lambda_j + \theta_j^i = 0.$$

If we choose prices $p = \lambda$, and $\mu = \theta$, the solution to the planners problem coincides with the solution of each agents problem. Thus, not only is there an equilibrium price, but at that price the resulting trades lie in the core.

5.6 Application: principal–agent problem

The principal–agent problem involves an individual (Principal) that employs another (the Agent) to perform a task. The task is onerous, so the agent must be compensated for doing it. The difficulty is that the principal cannot observe directly if the agent has performed the task. What the principal can observe is a signal that is an imperfect indicator of the effort expended by the agent.

A principal who hires another to sell their product faces just this problem. The number of purchase orders the agent brings in is an imperfect signal of the effort they have exerted to hawk the principal's goods. A low volume of orders could be the result of laziness on the part of the agent or competitive factors beyond the agents control, for example, a recession or the introduction by a rival firm of a superior offering. A high volume of orders could be the result of hard work or just plain luck.

The principals problem is to determine a contract that will give the agent the incentive to exert the desired level of effort. Since the principal cannot observe the level of effort directly, the payments (or penalties) in the contract can depend only on the observed signal. For the problem to be non-trivial, principal and agent must have different attitudes to risk.

It is usual to assume that the principal is risk neutral, that is, cares only about expected monetary payoff. The idea is that the principle is usually large and well diversified. If the agent is also risk neutral, then the principal can solve the incentive problem by selling the 'firm' to the agent outright. It is usual to assume that the agent is risk averse. This is modeled by endowing the agent with a concave utility function.

We set up the principal–agent problem in the following way:

1. $A = \{a_1, a_2, \ldots, a_n\}$ a finite set of possible actions that the agent can take.
2. $S = \{s_1, s_2, \ldots, s_m\}$ is the set of possible signals that the principal can observe.
3. Let $p(s_i | a_j)$ be the probability of observing signal s_i given action a_j was undertaken by the agent.
4. Assume $p(s_i | a_j) > 0$ for each s_i, a_j.
5. The agent's disutility from undertaking action a_j is $d(a_j)$.
6. The agent's utility as a function of the wage ω and the action $a \in A$ is $u(\omega) - d(a)$.
7. Assume u strictly increasing, concave, continuous and differentiable.
8. To model the fact that the agent is not obliged to accept any contract, the agent obtains a reservation utility of U_0.

The principal's problem is to determine the cheapest contract to induce the agent to adopt a given action, a_n, say. A contract is specified by stipulating a wage to be paid for each possible signal that is realized. Let $\omega(s_i)$ be the wage paid if signal s_i is realized.

It is natural to formulate the principals problem with the $\omega(s_i)$'s as the variables. However, this leads to an optimization problem with non-linear constraints. To avoid this we make a change of variables. If signal s_i is realized, the agent's utility will be $z_i = u(\omega(s_i))$. Our variables will be the z_i's. In words we formulate the problem in terms of the utility delivered to the agent rather than wage.

Given the z_i's we can determine the corresponding wage by inverting u. Since u is a strictly increasing function of ω it has an inverse v, that is, $\omega(s_i) = v(z_i)$, where $v = u^{-1}$. In addition, because u is concave, v is convex.

To induce the agent to undertake action a_n, the z_i's must be chosen so that

$$\sum_{i=1}^{m} p(s_i | a_n) z_i - d(a_n) \geq \sum_{i=1}^{m} p(s_i | a_j) z_i - d(a_j) \quad \forall a_j \neq a_n.$$

This is called an incentive compatibility constraint. To induce the agent to accept the contract

$$\sum_{i=1}^{m} p(s_i | a_n) z_i - d(a_n) \geq U_0.$$

This is called the individual rationality constraint.
The principal's optimization problem is:

$$\min \sum_{i=1}^{m} p(s_i|a_n)v(z_i)$$

$$\text{s.t.} \quad \sum_{i=1}^{m} p(s_i|a_n)z_i - d(a_n) \geq \sum_{i=1}^{m} p(s_i|a_j)z_i - d(a_j), \quad \forall a_j \neq a_n,$$

$$\sum 1_{i=1}^{m} p(s_i|a_n)z_i - d(a_n) \geq U_0.$$

The constraints are linear, the objective is to minimize a convex function, so the problem is an instance of a concave programming problem. For this constrained minimization problem the Kuhn–Tucker–Karush conditions yield

$$v'(z_i) = \lambda + \sum_{j=1}^{n} \mu_j \left(1 - \frac{p(s_i|a_j)}{p(s_i|a_n)} \right).$$

Here λ is the multiplier associated with the individual rationality constraint and $\{\mu_j\}$ the multiplers associated with the incentive compatability constraints. Since v is a convex function, its first derivative is an increasing function in z. The right side of the equation is composed of a fixed component λ and terms of the form $p(s_i|a_j)/p(s_i|a_n)$ that depend on the data. This ratio measures the likelihood action a_j was taken rather than a_n when signal s_i was observed. When these ratios are small for all $a_j \neq a_n$, it means that signal s_i is a strong indicator that action a_n was taken. Now compare two signals, s_i and s_t and suppose that s_i is a stronger indicator than s_t that action a_n was taken.[9] Then, by the first order condition we have $v'(z_i) \geq v'(z_t)$. So that wages should be higher for more informative signals than for less informative ones.

Problems

5.1 Consider the feasible region defined by $\{x_1 \geq 0, x_2 \geq 0, x_2 - (x_1 - 1)^2 \leq 0\}$. The point $(1, 0)$ is feasible. Would the Kuhn–Tucker theorem apply to this point?

5.2 Consider the following optimization problem:

$$\max\{x^2 - y^2 : x^2 + y^2 \leq 1\}$$

Write down the necessary conditions of Theorem 5.6 for a local maxima. Find all solutions to the system of equations you generate. Is the constraint qualification satisfied for each of them? Are all of them local maxima?

5.3 Consider the following optimization problem:

$$\max\{x^2 - y^2 : x^2 + y^2 = 1\}$$

Write down the necessary conditions for a local maxima. Find all four solutions to the system of equations you generate. Is the constraint qualification satisfied for each of them? Are all of them local maxima?

5.4 Solve the following problem:

$$\max\{\log x + \log y : x^2 + y^2 = 1\}$$

5.5 Decide which, if any, of the following functions is convex, concave or neither on the reals:

1. $2x^3 - 3x^2$
2. $xy - x^2 - y^2$
3. $3x + 2x^2 + 4y + y^2 - 2xy$
4. $x^2 + 3xy + 2y^2$
5. xy

5.6 Let $f(x) = \sum_{j=1}^{n} \alpha_j x_j^{\theta_j}$ where $\theta_j \neq 0$ for all j, $\theta_j \leq 1$ for all j and each α_j has the same sign as θ_j. Show that f is concave on the non-negative orthant.

5.7 Let A be convex subset of $\mathbb{R}^n$ and $B \subset \mathbb{R}^n$ (not necessarily convex). Suppose for each $c \in A$, the problem $f(c) = \min\{cx : x \in B\}$ has a solution. Show that $f(c)$ is concave on A.

5.8 Show that a differentiable real valued function on $\mathbb{R}$ is concave iff $f(x+a) \leq f(x) + af'(x)$ for all x, a.

5.9 Let C be convex and $f : C \to \mathbb{R}$. f is called **strictly concave** if $f(\lambda x + (1-\lambda y)) > \lambda f(x) + (1-\lambda)f(y)$ for all $x, y \in C$. If f is strictly concave show that arg max$\{f(x) : x \in C\}$ is either empty or unique.

5.10 A real valued function f on a convex set $C \subset \mathbb{R}^n$ is called **quasi-concave** if for all $x, y \in C$

$$f(\lambda x + (1-\lambda)y) \geq \min[f(x), f(y)].$$

Prove the following facts about quasi-concave functions:

1. The set $F_t = \{x \in C : f(x) \geq t\}$ for each real number t is convex.
2. The minimum of two quasi-concave functions is quasi-concave.
3. Any positive multiple of a quasi-concave function is quasi-concave.
4. Is the sum of two quasi-concave functions quasi-concave? Prove or give a counter-example.
5. Is $f(x) = x^3$ quasi-concave for all real numbers x?
6. Is a local maximum of a quasi-concave function a global maximum? Prove or give a counter-example.

7. A real valued function f on a convex set $C \subset \mathbb{R}^n$ is called **strictly quasi-concave** if for all $x, y \in C$

$$f(\lambda x + (1 - \lambda)y) > \min[f(x), f(y)].$$

Show that every local maximum of a strictly quasi-concave function is a global maximum.

8. Let f be a continuous and differentiable function on a convex set C. For all $x, y \in C$ with $x \neq y$ show that f is quasi-concave iff $f(y) \geq f(x) \to \nabla f(x) \cdot (y - x) \geq 0$.

9. Show that the Cobb–Douglas production function $x^p y^q$ is concave on the non-negative orthant if $p + q \leq 1$ but quasi-concave if $p + q > 1$.

5.11 Let α and β be positive constants. Show that the function $f(x) = -\alpha \sqrt{\delta^2 - x^2}$ is convex for $-\delta \leq x \leq \delta$. Use this result to prove that the set

$$\left\{ (x, y) \in \mathbb{R}^2 \colon \left(\frac{x}{a} \right)^2 + \left(\frac{y}{b} \right)^2 \leq 1 \right\}$$

(where $a, b > 0$) is convex.

5.12 Let $f \colon \mathbb{R}^n \to \mathbb{R}$ be convex and $g \colon \mathbb{R}^n \to \mathbb{R}$ be concave. Suppose $g(x) \leq f(x)$ for all $x \in \mathbb{R}^n$. Use the separating hyperplane theorem to show that there exists $c \in \mathbb{R}^n$ and $t \in \mathbb{R}$ such that

$$g(x) \geq c \cdot x + t \geq f(x), \quad \forall x \in \mathbb{R}^n.$$

5.13 Show that the function $f(x) = \log x$ is concave. Use this fact to prove that for any collection $\{a_1, a_2, \ldots, a_n\}$ of non-negative numbers

$$\frac{\sum_{j=1}^n a_j}{n} \geq \left[\prod_{j=1}^n a_j \right]^{1/n},$$

5.14 Show that $f(x) = x^2$ is convex for $x \geq 0$. Use that to prove the Cauchy–Schwarz inequality:

$$\sum_{i=1}^n x_i y_i \leq \left[\sum_{i=1}^n x_i^2 \right]^{1/2} \left[\sum_{i=1}^n y_i^2 \right]^{1/2},$$

where $x_i, y_i > 0$ for all i.

5.15   Prove the generalization of the Cauchy–Schwarz inequality called Hölders inequality:

$$\sum_{i=1}^{n} x_i y_i \leq \left[\sum_{i=1}^{n} x_i^p\right]^{1/p} \left[\sum_{i=1}^{n} y_i^q\right]^{1/q},$$

where x_i, $y_i > 0$ for all i and $p > 1$, $q > 0$ and $1/p + 1/q = 1$.

5.16   Deduce from Hölder's inequality, Minkowski's inequality:

$$\left[\sum_{i=1}^{n} (x_i + y_i)^p\right]^{1/p} \leq \left[\sum_{i=1}^{n} x_i^p\right]^{1/p} + \left[\sum_{i=1}^{n} y_i^q\right]^{1/q},$$

where x_i, $y_i > 0$ for all i and $p \geq 1$.

5.17   Solve max $2x_1 x_2 + 2x_2 - x_1^2 - 2x_2^2$.

5.18   Solve:

max $15x_1 + 30x_2 + 4x_1 x_2 - 2x_1^2 - 4x_2^2$

s.t. $x_1 + 2x_2 \leq 30$,

$x_1, x_2 \geq 0$.

5.19   Prove using KTK that $x = 2/\sqrt{3}$ and $y = 1.5$ is an optimal solution to

max $4x + 6y - x^3 - 2y^2$

s.t. $x + 3y \leq 8$,

$5x + 2y \leq 14$,

$x, y \geq 0$.

5.20   Let A be a $m \times n$ matrix, C a $n \times n$ matrix and x^* an optimal solution to the following program:

$$\min \left\{ \tfrac{1}{2} xCx + px : Ax \geq b \right\}.$$

Denote the ith row of A by a^i. Let $I = \{i : a^i x^* = b_i\}$. Show that there must be numbers $w_i \geq 0$ for all $i \in I$ such that

$$Cx^* + p = \sum_{i \in I} w_i a^i.$$

If C is positive semidefinite, show that the above necessary condition for optimality is also a sufficient condition.

5.21 For $x \geq 0$ define $f(x)$ as follows.

$$f(x) = \begin{cases} x \ln x, & x > 0, \\ 0, & x = 0. \end{cases}$$

Show that f is a convex function on $\mathbb{R}_+$. Prove that

$$xy \leq f(x) + e^{y-1}$$

for all $x \geq 0$ and $y \in \mathbb{R}$.

5.22 Let f be a real valued concave function on a compact convex subset C of $\mathbb{R}^n$. If f attains a minimum over C prove that it does so at one of the extreme points of C.

5.23 Consider the following quadratic program.

$$\min \ x^T Q x - bx$$

$$\text{s.t.} \quad Ax = c.$$

Prove that x^* is a local minimum iff it is a global minimum. Note that no convexity or concavity is assumed in the objective function.

5.24 Let M be a m by n matrix and $g \in \mathbb{R}^m$. Suppose there is no x such that $Mx = g$. Consider the least squares problem: $\min\{|Mx - g|^2 : x \in \mathbb{R}^n\}$. Show that x^* is an optimal solution to this problem iff $M^T M x = M^T g$. Prove that if the columns of M are LI, then the optimal solution is unique. Note $|Mx - g|$ is just the distance between Mx and g.

5.25 Consider the following optimization problem:

$$\min \ \tfrac{1}{2} x Q x + \tfrac{1}{2} y Q y - cx$$

$$\text{s.t.} \quad Ax + Qy \geq c,$$

$$x \geq 0.$$

Q is invertible, symmetric ($Q = Q^T$) and positive semi-definite ($u^T Q u \geq 0$ for all u) and A is skew symmetric ($A^T = -A$).[10]

Prove that the program is feasible, and has an optimal objective function value is 0.

Notes

1 Those interested in these matters should consult Chiang (1984) or Mas-Colell *et al.* (1995).
2 Other instances may be found in Milgrom and Segal (2002).
3 A principle celebrated in latin as '*de gustibus non est disputandum*'.
4 It makes its first appearance in the writings of the 18th century French physiocrat Anne Robert Jacques Turgot: The earth's fertility resembles a spring that is being pressed

downwards by the addition of successive weights. If the weight is small and the spring is not very flexible, the first attempts will have no results. But when the weight is enough to overcome the first resistance then it will give to the pressure. After yielding a certain amount it will again begin to resist the extra force put upon it, and weights that formerly would have caused a depression of an inch will now scarcely move it by a hair's breadth. And so the effect of additional weights will diminish.

5 It's been said of Edgeworth that he was 'adept at avoiding conversational English'. He once asked T. E. Lawrence (of Arabia) :'Was it very caliginous in the Metropolis?' Back came the reply: 'Somewhat caliginous but not altogether inspissated'.

6 The father of Economics.

7 Leacock (1936).

8 Assuming you have no control over the stock market.

9 One could formalize this by saying that $p(s_i|a_j)/p(s_i|a_n) < p(s_t|a_j)/p(s_t|a_n)$ for all j.

10 Skew-symmetry implies that $x^\mathrm{T} A x = 0$ for all x.

References

Chiang, A. C.: 1984, *Fundamental methods of mathematical economics*, 3rd edn, McGraw-Hill, New York.

Leacock, S.: 1936, *Hellements of hickonomics, in hiccoughs of verse done in our social planning mill*, Dodd, Mead & Company, New York.

Mangasarian, O. L.: 1994, *Nonlinear programming*, Classics in applied mathematics; 10, Society for Industrial and Applied Mathematics, Philadelphia.

Mas-Colell, A., Whinston, M. D. and Green, J. R.: 1995, *Microeconomic theory*, Oxford University Press, New York.

Milgrom, P. and Segal, I.: 2002, Envelope theorems for arbitrary choice sets, *Econometrica* **70**(2), 583–601.

6   Fixed points

The fixed point problem is this:

> Given a set $S \subset \mathbb{R}^n$ and a function $f: S \to S$, is there an $x \in$ such that $f(x) = x$?

The problem of finding the zeros of a function, f, i.e., an $x \in S$ such that $f(x) = 0$, can be converted into a fixed point problem. Observe that $f(x) = 0$ iff. $g(x) = x$ where $g(x) = f(x) + x$. Concave programming is also a special case of the fixed point problem. In the unconstrained case, the optimal solution is found by solving $\nabla f(x) = 0$. Conversely, the fixed point problem can be converted to the optimization problem $\min_{x \in S}(f(x) - x)^2$, but this is rarely helpful.

6.1   Banach fixed point theorem

The simplest of all fixed point theorems is ascribed to Stefan Banach (1892–1945).[1]

Definition 6.1   *A function $f: S \to S$ is called a **contraction mapping** if $d(f(x), f(y)) \leq \theta d(x, y)$ for all $x, y \in S$, where $0 \leq \theta < 1$ is a fixed constant.*

In the one-dimensional case, the contraction mapping condition is $|f(x) - f(y)| \leq \theta|x - y|$.

Theorem 6.2   *Let $S \subset \mathbb{R}^n$ be closed and $f: S \to S$ a contraction mapping. Then there exists a unique $x \in S$ such that $f(x) = x$.*

Proof   The proof is provided for the one-dimensional case. Choose any $x^0 \in S$ and let $x^n = f(x^{n-1})$. If $\{x^n\}_{n \geq 1}$ has a limit x^* then $x^* \in S$ because S is closed, and $f(x^*) = x^*$. Therefore, it suffices to prove that $\{x^n\}_{n \geq 1}$ has a limit. We use the Cauchy criterion.
Pick $q > p$. Then

$$|x^q - x^p| = \left| \sum_{n=p}^{q-1} (x^{n+1} - x^n) \right| \leq \sum_{n=p}^{q-1} |x^{n+1} - x^n|.$$

But

$$|x^{n+1} - x^n| = |f(x^n) - f(x^{n-1})| \leq \theta |x^n - x^{n-1}|.$$

Repeated application of the above yields

$$|x^{n+1} - x^n| \leq \theta^n |x_1 - x^0|.$$

Hence

$$|x^q - x^p| \leq \sum_{n=p}^{q-1} \theta^n |x^1 - x^0|$$

$$\leq |x^1 - x^0|(\theta^p + \theta^{p+1} + \cdots) = |x^1 - x^0| \frac{\theta^p}{1 - \theta}.$$

The last term goes to zero as p, q go to infinity because $\theta < 1$. Thus, $\{x^n\}_{n \geq 1}$ has a limit. We leave the proof of uniqueness as an exercise for the reader. ■

The Banach Theorem is quite weak. Consider $f: [0, 1] \rightarrow [0, 1]$, where $f(x) = x$. This function barely misses being a contraction since $|f(x) - f(y)| = |x - y|$ for all $x, y \in [0, 1]$. However, every point in $[0, 1]$ is a fixed point of this function.

The Banach theorem should really be interpreted as a sufficient condition for a certain simple algorithm to compute a fixed point.

6.2   Brouwer fixed point theorem

The big fixed point theorem is due to L. E. J. Brouwer (1881–1966).[2]

Theorem 6.3   *If $S \subset R^n$ is compact and convex and $f: S \rightarrow S$ is continuous there exists $x \in S$ such that $f(x) = x$.*

All of the assumptions in the theorem are essential. Suppose we drop only continuity. Consider $f: [0, 1] \rightarrow [0, 1]$ where $f(x) = 1$ when $x = 0$ and zero otherwise. There is clearly no fixed point in this case.

Now remove compactness alone. Let $S = \{x : x \geq 0\}$ and $f(x) = x + 1$. Clearly f has no fixed point.

Next relax just convexity. Let S be the boundary of the unit circle and f the function that rotates each point on the circle one degree to the right.

The one dimensional version of Brouwer's theorem is known as the intermediate value theorem: if a continuous function can take both positive and negative values then there must be a value where it is zero.[3] Here a proof for the one-dimensional version that is a little more involved than usual is presented, but has the advantage that it can be generalized.

Lemma 6.4   *Let $f: [0, 1] \to [0, 1]$ be continuous. Then there is an $x \in [0, 1]$ such that $f(x) = x$.*

Proof   Each $p \in [0, 1]$ can be represented as a convex combination of the end points of the interval:

$$p = (1 - p) \times 0 + p \times 1.$$

The same will be true for $f(p)$. So we express each $p \in [0, 1]$ as a pair of non-negative numbers $(p_1, p_2) = (1 - p, p)$ that add to one. When expressing $f(p)$ in this way we will write it as $(f_1(p), f_2(p)) = (1 - f(p), f(p))$. Suppose for a contradiction, that f has no fixed point.

Since $f: [0, 1] \to [0, 1]$ we can think of the function f as moving each point $p \in [0, 1]$ either to the right (if $f(p) > p$) or to the left (if $f(p) < p$). The assumption that f has no fixed point eliminates the possibility that f leaves the position of p unchanged.

Given any $p \in [0, 1]$ we label it (or color it) with a '(+)' if $f_1(p) < p_1$ (move to the right) and color it '(−)' if $f_1(p) > p_1$ (move to the left).[4] The assumption of no fixed point implies $f_1(p) \neq 1 - p$ for all $p \in [0, 1]$. Thus the labeling scheme is well defined.

Notice that the point 0 will be labeled (+) and the point 1 will be labeled (−).

Now choose any finite partition, Π^0, of the interval $[0, 1]$ into smaller intervals. This partition must contain a sub-interval $[p^0, q^0]$ whose endpoints have different labels. Here is why. Every endpoint of these subintervals is labeled either (+) or (−). The point '0', which must be the endpoint of one of the subintervals of Π^0 has label (+). The point '1' has label (−). As we travel from 0 to 1 (left to right) we leave a point labeled (+) and arrive at a point labeled (−). At some point, we must pass through a subinterval which has endpoints with different labels.

Now take the partition Π^0 and form a new partition Π^1, finer than the first by taking all the sub-intervals in Π^0 and cutting them in half. In Π^1 there must be at least one sub-interval, $[p^1, q^1]$ with endpoints having different labels. Repeat this procedure indefinitely.

This produces an infinite sequence of sub-intervals $\{[p^n, q^n]\}$ shrinking in size with different labels at the end points. Furthermore we can choose a subsequence of them so that the left hand end point, p^n, is labeled (+) and the right hand end point, q^n is labeled (−). Since these intervals live in $[0, 1]$ their lengths are bounded. Therefore by the Bolzano-Weierstrass theorem, there is a convergent subsequence of them, with $|p^n - q^n| \to 0$. By continuity, $f(p^n) - f(q^n) \to 0$.

Let r be the limit of p^n and q^n. By continuity, $f(p^n)$ and $f(q^n)$ both converge to $f(r)$. Since each p^n is labeled (+) and each q^n is labeled (−), for each n we have $f_1(p^n) < p_1^n$ and in the limit $f_1(r) \leq r_1$. For each n we have $f_1(q^n) > q_1^n$ and in the limit $f_1(r) \geq r_1$. Thus $f_1(r) \leq r_1$ and $f_1(r) \geq r_1$. This implies that $f_1(r) = r_1$ i.e., $f(r) = r$, we have a fixed point, a contradiction.   ∎

6.2.1 Sperner's lemma

To generalize the proof of Lemma 6.4 to n-dimensions, we need Sperner's Lemma. Before stating and proving it we need to discuss triangulations which are the natural generalization of the partition operation used in the proof of Lemma 6.4.

A **triangulation** of a triangle is a subdivision of the initial triangle into smaller triangles. There are many ways to triangulate a triangle, for our purposes one particular way will suffice.

Suppose we have a triangle, call it the 'big' triangle and label its three corners A, B and C. Now identify the mid-points of the line segments AB, BC and AC. In the big triangle draw a triangle whose three corners are the midpoints identified previously. Call this operation a subdivision. Notice that the subdivision divides the big triangle into 4 smaller triangles. This is illustrated in Figure 6.1. Call this triangulation the first subdivision of the big triangle. If we apply the subdivision operation to each of the four smaller triangles we obtain another triangulation, finer than the first, consisting of 16 smaller triangles. Call this triangulation the second subdivision. This is shown in Figure 6.2. If we apply the subdivision operation to each of the smaller triangles in the second subdivision we call the resulting triangulation the third subdivision and so on.

In the sequel we will apply this subdivision operation repeatedly to generate a finer and finer triangulation such that the length of the longest side of each of the small triangles goes to zero. This is not true of every triangulation. However, if our big triangle is an equilateral one, than it is true for the triangulation produced by the subdivision process described above.

It is usual to refer to the smaller triangles produced from the subdivision as **cells**. A boundary between two cells is called a **face**. The corners of the triangles produced by the subdivision are called **vertices**.

We now state and prove the two dimensional version of Sperner's Lemma. The n-dimensional version is essentially identical but consumes more notation.

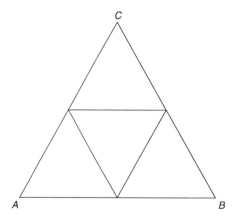

Figure 6.1

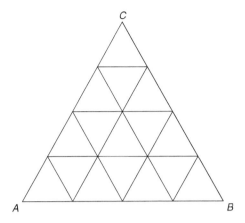

Figure 6.2

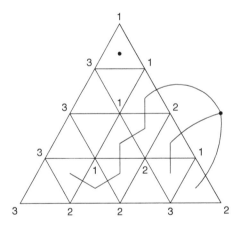

Figure 6.3

Theorem 6.5 (Sperner's lemma) *Let T be a triangle whose corners are V_1, V_2 and V_3. Let $\{T^1, T^2, \ldots, T^k\}$ be any triangulation of T. Suppose V_i gets the color i; and any vertex on the edge $[V_i, V_j]$ gets colored i or j. Then there exists a T^i whose three corners get three distinct colors.*

Proof We associate a graph with the triangulation. To each triangle in $T \cup \{T^1, T^2, \ldots, T^k\}$ associate a vertex. If T^r and T^q share a side one of whose endpoints is colored 1 and the other 2, put an edge between the corresponding vertices. In this graph, there will be an even number of odd degree vertices. An example of such graph is shown in Figure 6.3.

In particular the vertex associated with T will have odd degree. This is because the edge between V_1 and V_2 will have an odd number of color changes from 1 to 2. Therefore there are an odd number of vertices from $\{T^1, T^2, \ldots, T^k\}$ with odd degree. Each one of the smaller triangles with odd degree must be tri-colored. ∎

6.2.2 Application: cake cutting

A 'cake' corresponding to the interval $[0, 1]$, must be divided amongst n people by cutting it into n sub-intervals.[5] Denote the size of the ith piece by x_i. Notice that

$$\sum_{i=1}^{n} x_i = 1, \ x_i \geq 0, \quad \forall i.$$

Call a **cut**, any feasible solution to the above system. A **division** of the cake consists of a cut and an assignment of pieces to individuals. A division is **envy-free** if each player prefers their assigned piece to any other piece.

Two assumptions on preferences are required:

1. each player prefers something to nothing;
2. any piece that is preferred for a convergent sequence of cuts is preferred at the limiting cut set. If $\{x^n\}$ is a convergent sequence of cuts with limit x, and if for each n an agent prefers x_i^n to all other pieces, she will prefer x_i to all other pieces.

Do envy-free division's exist? We use Sperner's lemma to show they do. The argument will be carried out for the case $n = 3$, but can easily be generalized. Call the three agents A, B and C.

The feasible set of cuts is the set $T = \{x \in \mathbb{R}^3 : x_1 + x_2 + x_3 = 1, \ x_i \geq 0\}$ which is an equilateral triangle in a three dimensional space. The corners of this triangle have the coordinates $(1, 0, 0), (0, 1, 0)$ and $(0, 0, 1)$. We will triangulate this set using the subdivision operation described in Section 6.2.1. What will also matter is how we label each of the endpoints of the cells formed.

First label the endpoints of the triangle T, $A = (1, 0, 0)$, $B = (0, 1, 0)$ and $C = (0, 0, 1)$. Now form the first subdivision of the triangle ABC. The midpoint of the segment AB is labeled C, the midpoint of the segment AC is labeled B and the midpoint of the segment BC is labeled A. The labeling rule we use is this: the midpoint of a segment is labeled with a letter different from the labels on the end points that define it. Now form the second subdivision and apply the labeling rule to the midpoints used to construct the second sub-division and so on.

Consider the kth subdivision of T. Each labeled point corresponds to a cut of the cake. If the point is labeled A (B or C), ask player A (B or C) which of the three pieces of the cake she would prefer. If she answers piece i, color that labeled point i. Do this for all labeled points. We show that this coloring satisfies the conditions of Sperner's lemma.

Observe that the corners of T will be colored 1, 2 and 3 respectively by the first assumption about preferences. The labeled points on the edge of T that joins $(1,0,0)$ to $(0,1,0)$ can never be labeled 3 by assumption 1. All such points are convex combinations of $(1,0,0)$ and $(0,1,0)$ which means that in all of them the third piece has size zero, so no agent would choose the third piece. Similarly with the other two edges of T.

Thus, from Sperner's lemma we conclude that there is a triangle in the kth subdivision of T that is tri-colored. Pick one of them and call this triangle (a^k, b^k, c^k). Now let $k \to \infty$. Consider the infinite sequence $\{(a^k, b^k, c^k)\}$. From this sequence pick out a subsequence k_m where a^{k_m} is colored 1, b^{k_m} is colored 2 and c^{k_m} is colored 3 for all m. Actually all that matters is a subsequence where corners with the same label across the sequence have the same color. This is always possible since the sequence is infinite and there are only three colors.

Now the sequence $(a^{k_m}, b^{k_m}, c^{k_m})$ may not be convergent, but since it resides in a compact set, it has, by the Bolzano-Weierstrass theorem, a convergent subsequence. So we may assume that $a^{k_m} \to p$ as $m \to \infty$. Since the triangles in the subdivisions have diameters that shrink, $b^{k_m}, c^{k_m} \to p$ as well.

On the sequence of cuts $\{a^{k_m}\}$, person A always claims the first piece. On the sequence of cuts $\{b^{k_m}\}$, person B always claims the second piece. On the sequence of cuts $\{c^{k_m}\}$, person C always claims the third piece. So, by the second assumption, on the cut p, A prefers the first piece, B the second piece and C the third. Thus p is our envy free allocation.

6.2.3   Proof of Brouwer's theorem

Definition 6.6   *The **n-simplex** is the set* $\Delta^n = \{x \in \mathbb{R}^n : \sum_{i=1}^{n} x_i = 1, x_i \geq 0, \forall i\}$

From the definition we see that Δ^n is convex and compact. We also can see that it is an $(n-1)$-dimensional object.

Lemma 6.7   *Let $f : \Delta^n \to \Delta^n$ be a continuous function. Then there exists $x \in \Delta^n$ such that $f(x) = x$.*

Proof   The case $n = 2$ is covered by Lemma 6.4. Here we provide a proof for the case $n = 3$. The case for higher values of n goes in a similar way. As in the proof of Lemma 6.4, we suppose that f has no fixed point. Let $f_i(x)$ denote the ith coordinate of $f(x)$.

For the case $n = 3$, Δ^3 is a two-dimensional triangle. Subdivide this triangle into smaller triangles using the subdivision procedure described in Section 6.2.1. Consider the mth subdivision. Color those points that are vertices in the mth subdivision according to the following rule: $c(x) = \min\{i: f_i(x) < x_i\}$. This rule is well defined as long as f has no fixed point. If not, there must be an $x \in \Delta^3$ such that $f_i(x) \geq x_i$ for all i. Since $x \in \Delta^3$ and $f(x) \in \Delta^3$ it follows that $\sum_{i=1}^{3} x_i = 1 = \sum_{i=1}^{3} f_i(x)$, i.e., $f_i(x) = x_i$ which is a contradiction since we assumed that f has no fixed point.

We show that this coloring rule satisfies the assumptions of Sperner's lemma. Observe first that $c(1,0,0) = 1$, $c(0,1,0) = 2$ and $c(0,0,1) = 3$. Now examine a point on the edge of Δ^3. Consider, for example, a point x on the edge joining $(1,0,0)$ to $(0,1,0)$. Notice that $x = \lambda(1,0,0) + (1-\lambda)(0,1,0) = (\lambda, 1-\lambda, 0)$ for suitable λ. By the coloring rule we deduce that $c(\lambda, 1-\lambda, 0) = 1$ or 2. So, points on the boundary of Δ^3 are colored in accordance with the requirements of Sperner's lemma.

Invoking Sperner's lemma we deduce that in the mth subdivision of Δ^3 there exists a triangle with corners $(e^{m:1}, e^{m:2}, e^{m:3})$ that is tri-colored. Furthermore, we may, without loss of generality, assume that $c(e^{m:1}) = 1$, $c(e^{m:2}) = 2$ and $c(e^{m:3}) = 3$.

Let $m \to \infty$ and consider the sequence $\{e^{m:1}\}_{m \geq 1}$. It may not have a limit, but since it belongs to a compact set, it has, by the Bolzano–Weierstrass theorem, a convergent subsequence. So, for an appropriate subsequence, we may suppose $e^{m:1} \to x \in \Delta^3$. We also have $\{e^{m:2}\}, \{e^{m:3}\} \to x$ since as $m \to \infty$ the successive cells shrink in size. By continuity of f, $f(e^{m:1}) \to f(x)$, $f(e^{m:2}) \to f(x)$, $f(e^{m:3}) \to f(x)$. But $f_1(e^{m:1}) < e_1^{m:1}$ which implies $f_1(x) \leq x_1$. Similarly $f_2(x) \leq x_2$ and $f_3(x) \leq x_3$. Adding these inequalities yields

$$1 = \sum_{i=1}^{3} f_i(x) \leq \sum_{i=1}^{3} x_i = 1$$

which is possible only if $f_i(x) = x_i$ for all possible values of i, i.e., x is a fixed point, a contradiction. ∎

We are not done yet, since Brouwer's theorem holds for any compact, convex set not just simplices. To extend Lemma 6.7 we make use of the topological equivalence of compact, convex sets.

If S is a compact convex set of dimension $n - 1$, we know from Theorem 3.27 that there is a $g: S \to \Delta^n$ and $g^{-1}: \Delta^n \to S$ such that g and g^{-1} are continuous. Define $h: \Delta^n \to \Delta^n$ as follows:

$$h(x) = g[f(g^{-1}(x))].$$

Since h is continuous, by Lemma 6.7 it has a fixed point x^*. Therefore $h(x^*) = g[f(g^{-1}(x^*))] = x^*$. We have $f(g^{-1}(x^*)) = g^{-1}(x^*)$. So $g^{-1}(x^*)$ is a fixed point for f.

6.3 Application: Nash equilibrium

An *n* person (finite) game in strategic form is described by the following:

- The set N of players of cardinality n;
- The finite set S^i of strategies of player $i \in N$. Elements of the set S^i are called **pure strategies**;
- If player i chooses $s_i \in S^i$, the payoff to player $k \in N$ is $u^k(s_1, s_2, \ldots, s_n)$.

Definition 6.8 *An n-tuple of pure strategies,* $s_i \in S^i$ *is called a* **Nash**[6] **equilibrium in pure strategies** *if for all* $k \in N$:

$$u^k(s_1, s_2, s_{k-1}, s_k, s_{k+1}, \ldots, s_n) \geq u^k(s_1, s_2, s_{k-1}, x, s_{k+1}, \ldots, s_n), \quad \forall x \in S^k.$$

The *n*-tuple $(s_1, s_2, s_{k-1}, x, s_{k+1}, \ldots, s_n)$ is frequently abbreviated to (s^{-k}, x).

Example 22 *A two person strategic form game with two pure strategies for each player can be represented using a payoff matrix. One such example is shown below:*

2, 1	−1, −1
−1, −1	1, 2

This game has two pure strategy equilibria. One where the row player chooses row 1 and the column player chooses column 1. The second is where the players choose row 2 and column 2, respectively.

Example 23 *Not all games have an equilibrium in pure strategies as the following game illustrates:*

1, −1	0, 0
0, 0	1, −1

If we enlarge the notion of strategy to include randomized strategies we can ensure that every game has an equilibrium. For each S^i let Δ^i be the set of probability vectors over S^i. That is the tth component of $p \in S^i$ is the probability that strategy $s_t \in S^i$ is played. Thus, when we write p_i^k we mean that player k plays their pure strategy s_i^k with probability p_i^k. The elements of Δ^i are called **mixed strategies**. The expected payoff to player k when each player $i \in N$ plays the mixed strategy $p^i \in \Delta^i$ is denoted $u^k(p^1, \ldots, p^n)$ where

$$u^k(p^1, \ldots, p^n) = \sum_{s_1 \in S^1} \cdots \sum_{s_n \in S^n} u^k(s_{i_1}, s_{i_2}, \ldots, s_{i_n}) p_{i_1}^1, \ldots, p_{i_n}^n.$$

Notice that $u^k(p^1, \ldots, p^n)$ is continuous in the p's.

Definition 6.9   *An n-tuple of mixed strategies* $(p^1, \ldots, p^n)$ *is called a* **Nash equilibrium** *if for all* $k \in N$

$$u^k(p^1, \ldots, p^n) \geq u^k(p^{-k}, q), \quad \forall q \in \Delta^k.$$

By linearity of expectation, it is enough in the above definition, to consider q that put probability one on the pure strategies.

Example 24   *The game*

2, 1	−1, −1
−1, −1	1, 2

has one Nash equilibrium involving mixed strategies. The row player plays row 1 with probability 3/5 and row 2 with probability 2/5. The column player plays column 1 with probability 2/3 and column 2 with probability 3/5.

Theorem 6.10   *Every n person (finite) game in strategic form has a Nash equilibrium.*

Proof   Set $M = \Delta^1 \times \cdots \times \Delta^n$. Notice that M is compact and convex. We define a continuous function f from M into itself. Now each $p \in M$ is a vector with $\sum_{i \in N} |S^i|$ components. The component associated with player k and her pure strategy $s_i^k \in S^k$ will be written p_i^k. The corresponding component of $f(p)$ will be denoted $f_i^k(p)$. Define f as follows:

$$f_i^k(p) = \frac{p_i^k + [u^k(p^{-k}, s_i^k) - u^k(p), 0]^+}{\sum_{s_j^k \in S^k} [p_j^k + [u^k(p^{-k}, s_j^k) - u^k(p), 0]^+]}.$$

Here $[x, 0]^+ = \max(x, 0)$.

The choice of denominator in the definition of f ensures that f is well defined. This is because

$$\sum_{s_j^k \in S^k} [p_j^k + [u^k(p^{-k}, s_j^k) - u^k(p), 0]^+] \geq \sum_{s_j^k \in S^k} p_j^k = 1.$$

In addition, $\sum_i f_i^k(p) = 1$. Therefore f maps M into itself. Also, as the reader should verify, f is continuous. Thus by the Brouwer theorem there is a $p \in M$

such that $f(p) = p$. Hence

$$p_i^k = \frac{p_i^k + [u^k(p^{-k}, s_i^k) - u^k(p), 0]^+}{\sum_{s_j^k \in S^k}[p_j^k + [u^k(p^{-k}, s_j^k) - u^k(p), 0]^+]}.$$

If $\sum_{s_j^k \in S^k}[u^k(p^{-k}, s_j^k) - u^k(p), 0]^+ = 0$ for all $k \in N$, then $u^k(p^{-k}, s_j^k) - u^k(p) \leq 0$ for all $s_j^k \in S^k$ and $k \in N$, which is the definition of a Nash equilibrium and the proof is complete.

Suppose then there is an agent k such that $\sum_{s_j^k \in S^k}[u^k(p^{-k}, s_j^k) - u^k(p), 0]^+ > 0$. We show that

$$[u^k(p^{-k}, s_i^k) - u^k(p), 0]^+ > 0, \quad \forall s_i^k \in S^k \text{ s.t. } p_i^k > 0.$$

Suppose not. Then, for some $s_i^k \in S^k$ with $p_i^k > 0$ we have $[u^k(p^{-k}, s_i^k) - u^k(p), 0]^+ = 0$. From the fixed point property we have

$$p_i^k = \frac{p_i^k + [u^k(p^{-k}, s_i^k) - u^k(p), 0]^+}{\sum_{s_j^k \in S^k}[p_j^k + [u^k(p^{-k}, s_j^k) - u^k(p), 0]^+]}$$

$$= \frac{p_i^k}{1 + \sum_{s_j^k \in S^k}[u^k(p^{-k}, s_j^k) - u^k(p), 0]^+} < p_i^k$$

a contradiction. Hence

$$u^k(p^{-k}, s_i^k) > u^k(p), \quad \forall s_i^k \text{ s.t. } p_i^k > 0.$$

Thus

$$\sum_{s_i^k \in S^k} p_i^k u^k(p^{-k}, s_i^k) > \sum_{s_i^k \in S^k} p_i^k u^k(p) = u^k(p).$$

But the left-hand side of the above is also $u^k(p)$ and so we get a contradiction which completes the proof. ∎

6.4   Application: equilibrium in exchange economies

There are many different market arrangements one could choose to study, but perfectly competitive economies hold a special place in the economic imagination. The economists Kenneth Arrow and Frank Hahn[7] explain why:

> There is by now a long and fairly imposing line of economists from Adam Smith to the present who have sought to show that a decentralized economy motivated by self-interest and guided by price signals would be compatible

with a coherent disposition of economic resources that could be regarded, in a well defined sense, as superior to a large class of possible alternative dispositions. Moreover the price signals would operate in a way to establish this degree of coherence. It is important to understand how surprising this claim must be to anyone not exposed to the tradition. The immediate 'common sense' answer to the question 'What will an economy motivated by individual greed and controlled by a very large number of different agents look like?' is probably: There will be chaos. That quite a different answer has long been claimed true and has permeated the economic thinking of a large number of people who are in no way economists is itself sufficient ground for investigating it seriously. The proposition having been put forward and very seriously entertained, it is important to know not only whether it *is* true, but whether it *could* be true.

The usual idealization of competitive markets makes the following assumptions:

1. A finite number of agents (consumers) specified by their utilities, endowments and shares in the profit of each firm.
2. Each agent (consumers and firms) is aware of the price of every good.
3. The transaction costs of a sale, purchase, etc. are zero.
4. There is no uncertainty.
5. Agents can buy and sell as much and as little as they want at the going price. Their transactions do not affect the price. (They are price-takers.)
6. A finite number of commodities.
7. A finite number of firms that are specified by their input-output function.
8. All firms have technologies that exhibit decreasing returns to scale.

The question that motivates what is to come is this: in a perfectly competitive market is there a set of prices at which the customers' demands will balance the firms' outputs? Yes. To see why, imagine a world consisting of just one commodity, soma. Imagine that an auctioneer were

1. to announce a price p,
2. ask each consumer how much soma they would buy at that price, and
3. ask each firm how much soma they would produce at that price.

If the amounts submitted by the consumers matched the amounts submitted by the firms, end of story. We have found the price we are looking for. Suppose no match. Say, the total amount demanded by the consumers is more than what the firms are willing to supply at the posted price. What should the auctioneer do? As the selling price of soma rises, we expect the amount of soma demanded by each customer will *decrease*. In parallel, each firm will respond by *increasing* its output of soma. If we are lucky (and we are), if we raise the price by just enough we will match demand with supply.

The important point is this: under the right price structure, agents acting *independently* to maximize their utility will specify demands so that supply exactly balances demand. The equilibrium prices along with the resulting allocation is called a **Walrasian Equilibrium** in honor of Leon Walras (1834–1910)[8] who conceived both the idealization of markets and the notion of equilibrium.

The justification that Walras offered for such a study is eloquent: "How could these economists' prove that the results of free competition were beneficial and advantageous if they did not just what these results were? And how could they know these results when they had neither framed definitions nor formulated relevant laws to prove their point?" In response to the criticism that the model was to spare to be relevant, Walras wrote:

> What physicist would deliberately pick cloudy weather for astronomical observation instead of taking advantage of a cloudless night?

His contemporaries did not share his vision.[9]

The set up has a finite set A of m agents each with an endowment $w^i \in \mathbb{R}^n_+$. We assume for simplicity that $w^i \gg 0$ for all $i \in A$. For each agent i there is a utility function $U^i \colon \mathbb{R}^n_+ \to \mathbb{R}$ that is continuous, strictly concave and locally insatiable. A utility function U is called **locally insatiable** if for any $x \in \mathbb{R}^n_+$ there is a y in a neighborhood around x such that $U(y) > U(x)$. Let $M \geq \sum_{i \in A} w^i$. Assume the vector M is known to all agents.[10] We assume no production so as to keep the presentation uncluttered.

Definition 6.11 *An **equilibrium** is a price vector* $p \in \mathbb{R}^n_+$ *and an allocation* $X = (x^1, x^2, \ldots, x^m)$ *such that*

1. $x^i \in \mathbb{R}^n_+$,
2. $\sum_{i \in A} x^i = \sum_{i \in A} w^i$,
3. $x^i \in \arg\max\{U^i(x) \colon px \leq pw^i, \ x \leq M, \ x \geq 0\}$.

We will use Brouwer's fixed point theorem to establish the existence of an equilibrium.

For each $i \in A$ let

$$d^i(p) = \arg\max\{U^i(x) \colon px^i \leq pw^i, \ x^i \leq M, \ x^i \geq 0\}.$$

Imposing the restraint $x^i \leq M$ makes the feasible region of the above optimization problem compact. The constraint 'bites' when one or more components of p is zero. Continuity and strict concavity of U^i implies that $d^i(p)$ is well defined and single valued for each $p \geq 0$.[11] Further, the constraint $x^i \leq M$ ensures that $d^i(p)$ is always bounded. Three observations are useful.

> **Observation 1**: In an optimal solution to $\max\{U^i(x) \colon px^i \leq pw^i, \ x \leq M, x \geq 0\}$, the budget constraint $px \leq pw^i$ will be binding.
>
> Let x^* be the optimal solution and suppose $px^* < pw^i$. By the assumption of local insatiability there is an x' in a neighborhood of x^* such that $px' \leq pw^i$ and $U^i(x') > U^i(x^*)$ contradicting the optimality of x^*.

Observation 2: $d^i(\mu p) = d^i(p)$ for all $\mu > 0$.

The feasible region of the optimization problem $\max\{U^i(x): px^i \leq pw^i, x \leq M, x \geq 0\}$ does not change when both the left and right hand side of the only constraint are scaled by the same positive amount.

Observation 3: $U^i(x) > U^i(d^i(p)) \Rightarrow px > pd^i(p)$.

A bundle x that generates more utility than the maximum possible subject to feasibility must be infeasible.

In view of Observation 2, we can restrict p to being in the simplex Δ^n.

The problem of finding an equilibrium price p and allocation X reduces to finding a price vector p that solves $E(p) = 0$ where $E(p) = \sum_{i=1}^m d^i(p) - \sum_{i=1}^m w^i$. Recall that such a problem can be solved by identifying a fixed point of $p + E(p)$. The hurdle that we must overcome is to show that the conditions of Brouwer's theorem are satisfied by this function. First we require that $E(p)$ be continuous. Second, if p 'lives' in a compact convex set then so should $p + E(p)$.

Lemma 6.12 *Let $\{p^t\}_{t\geq 1}$ be a sequence of prices in Δ^n with limit p. Then $d^i(p^t) \to d^i(p)$.*

Proof Let $x(t) = d^i(p^t)$. Since $x(t) \in \{x \in \mathbb{R}^n_+ : x \leq M\}$ we can assume that that the sequence $x(t)$ has a limit x^*. By Observation 1, $px(t) = pw^i$ for all t. Hence $px^* = pw^i$. If $x^* = d^i(p)$ we are done, so suppose not.

Since $x^* \neq d^i(p)$ there is a z such that $pz = pw^i$ and $U^i(z) > U^i(x^*)$. Set $a_t = (p^t \cdot z)/(p^t \cdot w^i)$. The assumption that $w^i \gg 0$ for all i ensures that a_t is well defined for t sufficiently large. Notice that $a_t \to 1$ as $t \to \infty$. By continuity of U^i we have $U^i(a_t z) \to U^i(z)$. However, $p^t a_t z = p^t w^i$ so $U^i(a_t z) < U^i(x^t)$ for all t sufficiently large which contradicts the fact that $U^i(z) > U^i(x)$. ∎

Lemma 6.13 (Walras' law) *For all $p \in \mathbb{R}^n_+$, $p \cdot \left[\sum_{i \in A} d^i(p) - \sum_{i \in A} w^i \right] = 0$.*

Proof By Observation 1, $pd^i(p) = pw^i$. Now sum up these equations over agent i to obtain the result. ∎

To prove the existence of an equilibrium we define $f: \Delta^n \to \Delta^n$ as follows:

$$f_j(p) = \frac{p_j + [0, \sum_{i \in A}(d^i_j(p) - w^i_j)]^+}{\sum_{k=1}^n (p_k + [0, \sum_{i \in A}(d^i_k(p) - w^i_k)]^+)}.$$

The numerator is a cousin of $p + E(p)$ modified to ensure non-negativity for all p. The denominator is a scaling factor to ensure that the modified form of $p + E(p)$ 'lives' in a simplex.

From Lemma 6.12 we see that f is continuous. So, by the Brouwer theorem there is a $p \in \Delta^n$ such that

$$p_j = \frac{p_j + [0, \sum_{i \in A}(d_j^i(p) - w_j^i)]^+}{\sum_{k=1}^n (p_k + [0, \sum_{i \in A}(d_k^i(p) - w_k^i)]^+)}. \tag{6.1}$$

We have two cases to consider. First suppose there is a good j such that $p_j > 0$ and $[0, \sum_{i \in A}(d_j^i(p) - w_j^i)]^+ = 0$. Substituting into 6.1 yields

$$p_j = \frac{p_j}{\sum_{k=1}^n (p_k + [0, \sum_{i \in A}(d_k^i(p) - w_k^i)]^+)}.$$

This last equation holds only if $\sum_{k=1}^n [0, \sum_{i \in A}(d_k^i(p) - w_k^i)]^+) = 0$, i.e., $[0, (d_k^i(p) - w_k^i)]^+) = 0$ for all goods k. If for any good k we have $\sum_{i \in A}(d_k^i(p) - w_k^i) < 0$, this would violate Walras' law. So,

$$\sum_{i \in A}(d_k^i(p) - w_k^i) = 0$$

for all goods k which gives us our equilibrium.

Now suppose there is a good j such that $p_j > 0$ and $[0, \sum_{i \in A}(d_j^i(p) - w_j^i)]^+ > 0$. Substituting this into 6.1 and using the fact that $p \in \Delta^n$, yields

$$p_j = \frac{p_j + [0, \sum_{i \in A}(d_j^i(p) - w_j^i)]^+}{\sum_{k=1}^n (p_k + [0, \sum_{i \in A}(d_j^i(p) - w_j^i)]^+)}$$

$$\geq p_j + [0, \sum_{i \in A}(d_j^i(p) - w_j^i)]^+ > p_j,$$

a contradiction.

We now know that there is a price at which supply equals demand, so what? One can achieve the same balance by rationing demand and supply. What is special about the equilibrium allocation of commodities produced by a perfectly competitive market? To answer this question we need three definitions.

Call an allocation $X = (x^1, \ldots, x^m)$ **feasible** if $x^i \in \mathbb{R}_+^n$ for all $i \in A$ and $\sum_{i \in A} x^i = \sum_{i \in A} w^i$. An allocation X **Pareto dominates** an allocation Y if $U^i(x^i) \geq U^i(y^i)$ for all $i \in A$ with strict inequality for at least one agent. A feasible allocation X is called **Pareto optimal** if there is no feasible allocation that Pareto dominates it.[12]

The equilibrium allocation is always Pareto optimal. This conclusion goes under the name of the first welfare theorem of economics and is stated and proved below.

The first welfare theorem is the basis for statements of the following form: free markets generate an allocation of goods that cannot be improved upon.

One has to be careful here. The equilibrium allocation cannot be improved upon in a Pareto sense. This does not prevent the equilibrium allocation from being wildly inequitable. For example, consider a cake to be divided between two people. The allocation where one gets 99% of the cake and the other 1% is Pareto optimal. So is the 50-50 split. Indeed, all divisions are Pareto optimal.

Theorem 6.14   *If (p, X) is an equilibrium, then the allocation X is Pareto optimal.*

Proof   Suppose not. Then there is a feasible allocation $Z = (z^1, z^2, \ldots, z^m)$ such that $U^i(z^i) \geq U^i(x^i)$ for all $i \in A$ with strict inequality for agent k say.

By Observation 3, $p \cdot z^k > p \cdot x^k$. Hence $\sum_{i \in A} p \cdot z^i > \sum_{i \in A} p \cdot x^i$. By feasibility

$$\sum_{i \in A} p \cdot z^i = \sum_{i \in A} p \cdot w^i = \sum_{i \in A} p \cdot x^i,$$

a contradiction.   ∎

The next theorem (called the second welfare theorem) shows that any feasible Pareto optimal allocation is an equilibrium allocation for a suitable price vector.

Theorem 6.15   *Let $X = (x^1, \ldots, x^m)$ be a Pareto optimal allocation such that $x^i \gg 0$ for all $i \in A$. In an economy where X is the initial endowment there is a $p \in \mathbb{R}^n_+$ such that (p, X) is an equilibrium.*

Proof   Let S be the set of aggregate allocations not Pareto dominated by X, i.e.,

$$S = \left\{ y = \sum_{i \in A} y^i : U^i(y^i) \geq U^i(x^i), \quad \forall i \in A \right\}.$$

Fix an agent k and let

$$T^k = \left\{ z: z = y - \sum_{i \in A} x^i, \ y \in S, \ U^k(y^k) > U^k(x^k) \right\}.$$

One can interpret T^k to be the set of trades that have to be executed so as to shift the allocation from X to an allocation $Y = (y^1, \ldots, y^n)$ where agent k is strictly better off and no other agent is worse off in utility terms.

Concavity of U^i for all i implies that T^k is a convex set. Further T^k contains no vector z such that $z \ll 0$. If it did, there would be an allocation $Y = (y^1, \ldots, y^m)$ such that $y \in S$ and $y = z + \sum_{i \in A} x^i$. Since $z \ll 0$, Y would be a feasible allocation and violate the Pareto optimality of X.

Since T^k is disjoint from the strictly negative orthant, by the weak separating hyperplane theorem there is a $p \in \mathbb{R}^n$ such that $p \cdot z \geq 0$ for all $z \in T^k$. Notice we do not have strict inequality because T^k is not a closed set, in fact it may contain the origin.

Since p (weakly) separates T^k from the negative orthant, this implies that $p \geq 0$. Hence, for all $y \in S$ such that $U^k(y^k) > U^k(x^k)$ we have $p \cdot y \geq p \cdot x$. Now we show that $p \cdot y \geq p \cdot x$ for all $y \in S$.

Suppose not. Then there exists $v \in S$ such that $U^k(v^k) = U^k(x^k)$ and $p \cdot v < p \cdot x$. Fix a $y \in S$ such that $U^k(y^k) > U^k(x^k)$. Choose a $\mu \in (0, 1)$ and let $w(\mu) = \mu y + (1 - \mu)v$. Notice $w(\mu) \in S$. Strict concavity of U implies

$$U^k(\mu y^k + (1 - \mu)v^k) > U^k(x^k).$$

Hence $p \cdot w(\mu) \geq p \cdot x$. Let $\mu \to 0$. Then $w(\mu) \to v$, but for all μ sufficiently small $p \cdot w(\mu) \geq p \cdot x$ which contradicts $p \cdot v < p \cdot x$.

To complete the proof let we must show that

$$x^i = \arg\max\{U^i(h) : p \cdot h \leq p \cdot x^i, h \in \mathbb{R}^n_+, h \leq M\}, \quad \forall i \in A.$$

Suppose this is not true for some agent t, say. Let

$$d^t = \arg\max\{U^i(h) : p \cdot h \leq p \cdot x^i, h \in \mathbb{R}^n_+, h \leq M\}.$$

By assumption $d^t \neq x^t$. Sub-optimality of x^t implies $U^t(d^t) > U^t(x^t)$ and strict concavity implies $U^t(\mu d^t) > U^t(x^t)$ for some $\mu \in (0, 1)$.

Let X' be the allocation obtained from X by replacing x^t with μd^t. Observe that $\sum_{i \neq t} x^i + \mu d^t \in S$ and so $p \cdot [\sum_{i \neq t} x^i + \mu d^t] \geq p \cdot x$. However

$$p \cdot \left[\sum_{i \neq t} x^i + \mu d^t\right] = \sum_{i \neq t} p \cdot x^i + \mu p \cdot d^t(p) < p \cdot x$$

since by Observation 1 $p \cdot d^t < p \cdot x^t$. This contradiction proves the theorem. ■

6.5 Application: Hex

The game of Hex is played on a rhombus shaped board with hexagonal cells.[13] The standard size is an 11×11 board and is shown in Figure 6.4. Two players, Black and White, are assigned opposite edges of the board. The board is initially empty. Black and white, move alternately marking a chosen hexagonal cell with their color. The game is won when one player establishes an unbroken chain of his pieces connecting his sides of the board. The game was invented by the Danish poet and architect Piet Hein (1905–1996) in 1942 who called it 'polygon'. It was discovered anew by John Nash in 1948. Legend has it that the game was played on the tiles of one of the bathrooms at Princeton University. There the game was

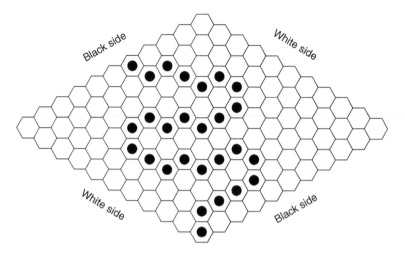

Figure 6.4

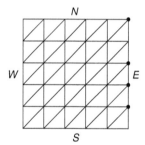

Figure 6.5

called Nash. It was sold commercially by Parker Brothers under the name Hex, but no longer.

We shall use Brouwer's theorem to prove that the game Hex can never end in a draw. We can model a $k \times k$ board as a graph with one vertex for each integral vector in $[1, k] \times [1, k]$. The set of such integral vectors we call B^k. Thus each cell of the $k \times k$ board corresponds to a point/vertex in the set B^k. Figure 6.5 illustrates B^5. Lines between vertices identify the adjacency relationships. Thus, the vertex $(x_1, x_2) \in B^k$ is adjacent to $(x_1 - 1, x_2)$, $(x_1, x_2 - 1)$, $(x_1 - 1, x_2 - 1)$, $(x_1 + 1, x_2)$, $(x_1, x_2 + 1)$ and $(x_1 + 1, x_2 + 1)$.

Vertices with coordinates $(1, \cdot)$, $(k, \cdot)$, $(k, \cdot)$ and $(\cdot, k)$ are boundary vertices. We label each set S, N, W and E, respectively. The vertex $(1, 1)$ counts as being on both the S and W border. The vertex (k, k) counts as being on the N and E border.

The 'horizontal' player seeks to mark vertices with an 'H' so that the marked vertices form a path from a vertex in E to a vertex in W. The 'vertical'player seeks to mark vertices with a '*V*', so that the marked vertices form a path from a vertex in *S* to a vertex in *N*.

Theorem 6.16 *Suppose each point of B^k is labeled either H (for horizontal) or V (for vertical). Then there is a path of H vertices between N and S or a path of V vertices between E and W but not both.*

Proof The 'not both' part of the theorem is left as an exercise. We suppose, for a contradiction that there is no path of *H* vertices between *N* and *S* and no path of *V* vertices between *E* and *W*.

In an abuse of notation we use *H* to denote the set of vertices labeled *H*, similarly with *V*. Let W' denote the set of vertices in *H* connected to a vertex in $H \cap W$ by a path consisting only of vertices in *H*. Let $E' = H \setminus W'$. No vertex in W' can be adjacent to a vertex in E'. Similarly let S' denote the vertices in *V* connected by a path consisting only of vertices in *V* to a vertex in $S \cap V$. Let $N' = V \setminus S'$. No vertex in S' can be adjacent to a vertex in N'. Note that $B^k = W' \cup E' \cup S' \cup N'$.

Define a function $f\colon B^k \to B^k$ as follows:

1. If $z \in W'$ then $f(z) = z + (1, 0)$.
2. If $z \in E'$ then $f(z) = z + (-1, 0)$.
3. If $z \in S'$ then $f(z) = z + (0, 1)$.
4. If $z \in N'$ then $f(z) = z + (0, -1)$.

We must verify that $f(z) \in B^k$. Suppose first that $z \in W'$ and $z+(1, 0) \notin B^k$. Then $z \in E$ which contradicts the initial assumption. The other cases follow similarly.

We now extend *f* in such a way that $f\colon [1, k] \times [1, k] \to [1, k] \times [1, k]$ and *f* is continuous. Pick any $x \in [1, k] \times [1, k]$. It is easy to see that there exists at most three pairwise adjacent vertices $z^1, z^2, z^3 \in B^k$ such that *x* lies in their convex hull. Hence we can set $x = \lambda_1 z^1 + \lambda_2 z^2 + \lambda_3 z^3$ where each λ_i is non-negative and $\lambda_1 + \lambda_2 + \lambda_3 = 1$. We will refer to z^1, z^2 and z^3 as the point *x*'s defining vertices. In this case define $f(x)$ to be $\lambda_1 f(z^1) + \lambda_2 f(z^2) + \lambda_3 f(z^3)$. It is easy to see that *f* defined in this way is continuous.

Since $[1, k] \times [1, k]$ is compact and convex, we can invoke the Brouwer theorem to conclude that there exists $x^* \in [1, k] \times [1, k]$ such that $f(x^*) = x^*$. Let z^1, z^2 and z^3 be the point x^*'s defining vertices. Then:

$$\lambda_1 z^1 + \lambda_2 z^2 + \lambda_3 z^3 = \lambda_1 f(z^1) + \lambda_2 f(z^2) + \lambda_3 f(z^3)$$
$$= \lambda_1 z^1 + \lambda_2 z^2 + \lambda_3 z^3 + \lambda_1 v^1 + \lambda_2 v^2 + \lambda_3 v^3$$
$$\Rightarrow \lambda_1 v^1 + \lambda_2 v^2 + \lambda_3 v^3 = 0$$

where $v^1, v^2, v^3 \in \{(1, 0), (-1, 0), (0, 1), (0, -1)\}$.

Suppose $z^1 \in W'$, a similar argument will apply in the other cases. Then $v^1 = (1, 0)$. Since z^2 and z^3 are adjacent to z^1 they cannot be in E'. Since z^2 and z^3 adjacent, either both are from S', both are from N', both from W' or one is from W' and the other from $S' \cup N'$. If $z^2, z^3 \in S'$ then $v^2 = v^3 = (0, 1)$, in which case $\lambda_1 v^1 + \lambda_2 v^2 + \lambda_3 v^3 = (\lambda_1, \lambda_2 + \lambda_3) \neq 0$ a contradiction since $\lambda_1 + \lambda_2 + \lambda_3 = 1$. If $z^2, z^3 \in N'$ then $v^2 = v^3 = (0, -1)$ in which case $\lambda_1 v^1 + \lambda_2 v^2 + \lambda_3 v^3 = (\lambda_1, -\lambda_2, -\lambda_3) \neq 0$ a contradiction. If $z^2, z^3 \in W'$ it is easy to see that $\lambda_1 v^1 + \lambda_2 v^2 + \lambda_3 v^3 = (1, 0, 0) \neq 0$ a contradiction. Similarly with the other cases.   ∎

6.6   Kakutani's[14] fixed point theorem

Definition 6.17   *A **correspondence** C on $S \subset \mathbb{R}^n$ is a rule that associates with each $x \in S$ a set $C(x) \subset S$.*

Definition 6.18   *A correspondence C is called **upper semi-continuous**, abbreviated to usc, if the set $\{(x, y): y \in C(x)\}$ is closed. The set $\{(x, y): y \in C(x)\}$ is called the **graph** of the correspondence.*

An equivalent definition of usc is to say that if $x^n \to x$ and $y^n \in C(x^n)$ for all n such that $y^n \to y$ then $y \in C(x)$.

Example 25   *Here is a correspondence defined on $[-1, 1]$. If $x \in [-1, 0)$ then $C(x) = 0.5$. If $x \in (0, 1]$ then $C(x) = -0.5$. If $x = 0$, $C(x) = \{0.5, -0.5\}$. It is easy to check that this correspondence is usc.*

The notion of usc generalizes the notion of continuity. It is easy to see that any continuous real valued function must be usc. The converse is not true. Consider $f(x) = x^{-1}$ for all $x \neq 0$ but $f(0) = 1$. The function is not continuous, but its graph is a closed set.

Definition 6.19   *A correspondence C on $S \subset \mathbb{R}^n$ is called **convex valued** if $C(x)$ is a convex set for all $x \in S$.*

The correspondence in the example above is not convex valued.

Theorem 6.20 (Kakutani's fixed point theorem)   *Let $S \subset \mathbb{R}^n$ be a compact and convex set. Let C be a correspondence from S into itself that is usc and convex valued. Then, there is an $x^* \in S$ such that $x^* \in C(x^*)$.*

Proof   As in the case of Brouwer's theorem, it suffices to prove the theorem for the case when $S = \Delta^3$.

Consider the mth subdivision of the simplex Δ^3. We associate with C a real valued function f^m in the following way.

If $x \in \Delta^3$ is a vertex of the mth subdivision, choose any $y \in C(x)$ and set $f^m(x) = y$. If $x \in \Delta^3$ is not a vertex of the mth subdivision of Δ^3, then x must be in some triangle/cell of the subdivision with corners/vertices e^1, e^2 and e^3, say. Further, x can be expressed as a convex combination of these three vectors, i.e. $x = \lambda_1 e^1 + \lambda_2 e^2 + \lambda_3 e^3$ where $\lambda_i \geq 0$ for all i and $\lambda_1 + \lambda_2 + \lambda_3 = 1$. In this case, set $f^m(x) = \lambda_1 f^m(e^1) + \lambda_2 f^m(e^2) + \lambda_3 f^m(e^3)$.

It is easy to see that $f^m: \Delta^3 \to \Delta^3$ is continuous given that C is usc. By Brouwer's theorem there exists $x^m \in \Delta^3$ such that $f^m(x^m) = x^m$. If x^m is a vertex of the mth subdivision, we are done because $x^m = f^m(x^m) \in C(x^m)$, by construction.

Suppose then that x^m is interior to one of the cells of the mth subdivision. Let $\{e^{m:1}, e^{m:2}, e^{m:3}\}$ be the corners of this triangle. Then $x^m = \lambda_1^m e^{m:1} + \lambda_2^m e^{m:2} + \lambda_3^m e^{m:3}$ where $\lambda_i^m \geq 0$ for all i and $\lambda_1^m + \lambda_2^m + \lambda_3^m = 1$. By our definition of f^m,

$$f^m(x^m) = \lambda_1^m y^{m:1} + \lambda_2^m y^{m:2} + \lambda_3^m y^{m:3},$$

where $y^{m:j} = f^m(e^{m:j})$. Since the sequences $\{x^m\}_{m \geq 1}$, $\{y^{m:j}\}_{m \geq 1}$ and $\{\lambda_j^m\}_{m \geq 1}$ for all j are all contained in a bounded set, it follows by the Bolzano-Weierstrass theorem that they all have a convergent subsequence. Let x^*, y^j and λ_j for all j be those limits.

Since the triangles of the subdivision are shrinking as $m \to \infty$, it follows that $e^{m:j} \to x^*$ for all j as $m \to \infty$. Since C is usc it follows that $y^j \in C(x^*)$. Also

$$x^* = \lambda_1 y^1 + \lambda_2 y^2 + \lambda_3 y^3.$$

Since $y^j \in C(x^*)$ for all j it follows from the convexity of C that $x^* \in C(x^*)$. This proves the theorem for the case when $S = \Delta^3$. ∎

Kakutani's theorem is useful in establishing the existence of equilibrium in exchange economies when utility functions are concave rather than strictly concave. Using the notation of Section 6.4, under concavity, $d^i(p)$ becomes a convex correspondence.

John Nash's original proof of the existence of equilibrium in games used Kakutani's theorem. We illustrate using a two person game and the notation of Section 6.3. Let S^i be the set of pure strategies for player i and let

$$B_1(q) = \arg\max \left\{ \sum_{i \in S^1} \sum_{j \in S^2} p_i q_j u^1(i, j) : \sum_{i \in S^1} p_i = 1, \ p_i \geq 0, \quad \forall i \in S^1 \right\}$$

and

$$B_2(p) = \arg\max \left\{ \sum_{i \in S^1} \sum_{j \in S^2} p_i q_j u^2(i, j) : \sum_{j \in S^2} q_j = 1, \ q_j \geq 0, \quad \forall j \in S^2 \right\}.$$

For each fixed q, $B_1(q)$ is the set of optimal solutions to a linear program. Similarly with $B_2(p)$. It is easy to see then $B_1(q)$ and $B_2(p)$ are convex correspondences.

Now define a correspondence C on $\Delta^1 \times \Delta^2$ as follows

$$C(p,q) = (B_1(q), B_2(p)).$$

It is easy to see that C satisfies all the conditions of Kakutani's theorem and the fixed point that is produced is a Nash equilibrium.

Problems

6.1 Show that the fixed point produced by the Banach theorem is unique.

6.2 Show that $f(x) = 1 - x^5$ is not a contraction mapping over $[0, 1]$ but that it does have a fixed point.

6.3 Let $f(x) = x + e^{-x}$ for $x \geq 0$. Is f a contraction mapping?

6.4 Let $f: \mathbb{R} \to \mathbb{R}$ be defined by $f(x) = 1/2(x + a/x)$ where a is a number strictly between 1 and 3. Is f a contraction mapping?

6.5 A function $f: S \to S$ with $S \subset \mathbb{R}^n$ closed is called **weakly contractive** if $d(f(x), f(y)) < d(x, y)$ $\forall x, y \in S$. Give an example of a weakly contractive mapping with no fixed point.

6.6 Consider $f: [0, 1] \to [0, 1]$ such that $f(x) = \sin x$. Show that f is weakly contractive but is not a contraction mapping.

6.7 Let $f: S \to S$ be weakly contractive and $S \subset \mathbb{R}^n$ compact. Show that f has a fixed point.

6.8 Let $C \subset \mathbb{R}^n$ be a non-empty, compact and convex set. A function $f: C \to C$ is called **affine** if $F(\lambda x + (1 - \lambda)y) = \lambda f(x) + (1 - \lambda)f(y)$ for all $x, y \in C$ and $\lambda \in [0, 1]$. Without appealing to the Brouwer theorem, give a short proof that every continuous and affine function f has a fixed point in C.

6.9 Let C be the boundary of a circle in $\mathbb{R}^2$ of finite radius. Let $f: C \to \mathbb{R}$ be continuous. Show that there are two points x^1 and x^2 on the circle C such that $f(x^1) = f(x^2)$ and the straight line joining them goes through the center of the circle.

6.10 Let $C \subset \mathbb{R}^n$ be a compact convex set and $f_i: C \to C$ for $i = 1, 2, \ldots, m$ a collection of continuous functions. Prove that there is an $x \in C$ such that

$$\sum_{i=1}^m f_i(x) = mx.$$

6.11 Let f be a continuous function that maps the letter 'Y' into itself. Show that there is a point on the letter Y that is fixed under the mapping. What other letters of the alphabet have such a fixed point property?

6.12 Let A be a $n \times n$ matrix with all entries strictly positive. Use Brouwer's theorem to show that there is a number $\lambda > 0$ and vector $x > 0$ such that $Ax = \lambda x$.

6.13 Let $S = \{v^0, v^1 v^2, \ldots, v^m\} \subset \mathbb{R}^{m+1}$ and $\{F_0, F_1, \ldots, F_m\}$ a collection of closed subsets of the convex hull of S such that for every $A \subset \{0, 1, \ldots, m\}$

we have

$$\text{conv}(\{v^i\}_{i\in A}) \subset \bigcup_{i\in A} F_i.$$

Use Brouwer's fixed point theorem to prove that $\bigcap_{i=0}^m F_i$ is compact and non-empty.

Hint: Define $g_i(x)$ to be the distance from x to the nearest point in F_i and $f_i(x) = (x_i + g_i(x))/(1 + \sum_{j=0}^m g_j(x))$. Apply Brouwer's theorem to f.

6.14 Let $\{f_1, \ldots, f_n\}$ be continuous functions from $\mathbb{R}^n$ to $\mathbb{R}$ with $f_i(x) > 0$ for all $i = 1, \ldots, n$ and all $x \in \mathbb{R}^n$. Show that there is an $x \in \mathbb{R}^n$ and $\lambda \in \mathbb{R}$ such that $x \geq 0$, $\sum_{j=1}^n x_j = 1$ and $f_i(x) = \lambda x_i$ for all i.

6.15 Find the Nash equilibria of the following games:

1.

4,4	0,5
5,0	1,1

2.

2,1	0,0
0,0	1,2

3.

10,10	10,6	10,1
6,10	14,14	8,2
1,10	2,8	10,10

6.16 Let C be a correspondence from $[0, 2]$ into itself defined as

$$C(x) = \begin{cases} 1, & 0 \leq x < 1, \\ [0, 2], & 1 \leq x \leq 2. \end{cases}$$

Show that C satisfies all the conditions of Kakutani's theorem.

6.17 Let C be a correspondence from $[0, 1]$ into itself defined as

$$C(x) = \begin{cases} x, & 0 \leq x < 1, \\ 0, & x = 1. \end{cases}$$

Is C upper-semi continuous?

6.18 For any $u, v \in \mathbb{R}^n$ with $u \leq v$ set $Q(u, v) = \{x \in \mathbb{R}^n : u \leq x \leq v\}$. Let $C \in \mathbb{R}^n$ be compact and convex and h and g continuous functions of C into itself such that $g(x) \leq h(x)$ and $g(x) \neq h(x)$ for all $x \in C$. Let f be a correspondence such that $f(x) = Q(g(x), h(x))$. Show that f is closed. Show also that f has a fixed point x such that $g(x) \neq x$ and $h(x) \neq x$.

6.19 Complete the proof of Kakutani's theorem by showing how to extend the result from a simplex to any compact convex set.

Notes

1 Part of a celebrated contingent of Polish mathematicians who would meet regularly at the Scottish Cafe to do mathematics. One of them, Stanislaw Ulam wrote this of Banach: 'It was difficult to outlast or outdrink Banach during these sessions. We discussed problems proposed right there, often with no solution evident even after several hours of thinking. The next day Banach was likely to appear with several small sheets of paper containing outlines of proofs he had completed'.
2 Notorious as the founder of the 'intuitionist' school in Mathematics. One of its tenets is the rejection of the proof by contradiction. In lectures Brouwer would never look at the students, only the blackboard and detested questions during class. The mathematician Van der Waerden who was a student in one of these classes writes: 'It seemed that he was no longer convinced of his results in topology because they were not correct from the point of view of intuitionism, and he judged everything he had done before, his greatest output, false according to his philosophy. He was a very strange person, crazy in love with his philosophy'.
3 Proved first by Bolzano.
4 Equivalently, p is colored $(+)$ if $f(p) > p$ and $(-)$ otherwise.
5 This section is based on Su (1999).
6 John Forbes Nash (1928–). Read the book, see the movie.
7 Arrow and Hahn (1971).
8 Rejected by the Ecole Polytechnique, he was to spend ten years as a mediocre journalist, bank clerk and railway official. Eventually he was awarded a chair in economics at the University of Lausanne.
9 'Since the world has won a victory over me, I am going to retire to a place of solitude where the world cannot reach me and where I can remain faithful to my dream'.
10 Standard treatments do not make this assumption. Dropping the assumption introduces some technical difficulties which we wish to avoid.
11 Strict concavity can be relaxed to concavity, but existence of equilibrium requires a different fixed point theorem that we discuss later.
12 The notion is due to Vilfredo Pareto (1848–1923) who succeeded Walras at Lausanne. Born an aristocrat he was a skilled swordsman and crack shot. Pareto once gave a talk where he was repeatedly interrupted by a German scholar, Gustav von Schmoller, who shouted that 'there are no laws in economics!' The next day, Pareto, his usual messy self, spied Schmoller in the streets. Pretending to be a beggar, Pareto approaches von Schmoller and says, 'Please, sir, can you tell me where I can find a restaurant where you can eat for nothing?' Schmoller replied, 'My dear man, there are no such restaurants.' 'Ah', said Pareto 'so there are laws in economics!'
13 This section is based on Gale (1979).
14 Shizuo Kakutani (1911–) father of the New York Times book reviewer Michiko Kaku-tani. A student once asked him if he could come to Kakutani's office at 4 p.m. that day. 'Yes', came the reply. 'And', continued the student, 'will you be there?'. 'No', was the response.

References

Arrow, K. J. and Hahn, F.: 1971, *General competitive analysis*, Mathematical economics texts, 6, Holden-Day, San Francisco.

Border, K. C.: 1985, *Fixed point theorems with applications to economics and game theory*, Cambridge University Press, Cambridge [Cambridgeshire], New York.

Gale, D.: 1979, The game of hex and the brouwer fixed-point theorem, *American Mathematical Monthly* **86**(10), 818–27.

Starr, R. M.: 1997, *General equilibrium theory: an introduction*, Cambridge University Press, Cambridge, New York.

Su, F. E.: 1999, Rental harmony: Sperner's lemma in fair division, *American Mathematical Monthly* **106**(10), 930.

7 Lattices and supermodularity

Many games of economic significance have the feature that the players have a continuum of strategies. For such games, existence of equilibrium cannot be deduced by an appeal to the Nash theorem. In these cases one must rely on properties of the payoff functions of the players. One such property is called supermodularity.

Existence of equilibria is not the only reason to be interested in supermodularity. Frequently one is interested in the behavior of a function as one changes some parameter. If the function is given explicitly in terms of the parameter this can be done using derivatives (assuming differentiability). However, in many cases the function one is studying is given indirectly. As an example, consider a firm facing a market price of y per unit of output. The cost to the firm of producing x units of output is $C(x)$. The firms profit as a function of output and market price is $f(x, y) = yx - C(x)$. Let the maximum possible profit as a function of price y be $g(y) = \max_{x \geq 0} f(x, y)$. Let the profit maximizing level of output be $x(y)$. Two natural questions are how $g(y)$ and $x(y)$ behave as the market price, y, changes. When the profit function $f(x, y)$ has the supermodularity property, it is possible to say just how $g(y)$ and $x(y)$ behave as y changes.

Recall the following notation to order vectors x and y in $\mathbb{R}^n$.

- $x = y$ iff $x_i = y_i$ for all i.
- $x \geq y$ iff $x_i \geq y_i$ for all i.
- $x > y$ iff $x_i \geq y_i$ for all i with strict inequality for at least one component.
- $x \gg y$ iff $x_i > y_i$ for all i.

We write $x \wedge y$ to mean the vector whose ith component is $\min\{x_i, y_i\}$. The vector $x \wedge y$ is sometimes called the **meet** of x and y. The vector $x \vee y$, called the **join**, is one whose ith component is $\max\{x_i, y_i\}$.

Definition 7.1 *A set $X \subset \mathbb{R}^n$ is called a **lattice** if for all $x, y \in X$ we have $x \wedge y$ and $x \vee y$ in X.*

Example 26 *The interval $[0, 1]$ is a lattice, the set $H = \{(x, y): x = y\}$ is a lattice as is the set $\{(1, 3), (4, 3), (3, 1), (1, 1)\}$. It is depicted in Figure 7.1. However the set $\{(x, y): x + y = 1\}$ is not.*

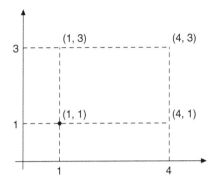

Figure 7.1

Example 27   *Let $X = \{x \in \mathbb{R}^n_+ : \sum_{i=1}^n x_i \leq 1\}$. Let e^i denote the vector with 1 in the ith component and zero elsewhere. Observe that e^i and e^k are both in X but $e^i \vee e^k = e^i + e^k \notin X$. So, X is not a lattice. However, it is possible to transform X into a lattice.*

Let $y_k = \sum_{i=1}^k x_i$ for $k = 1, \ldots, n$. Notice, $x_i = y_i - y_{i-1}$ for all $i = 1, \ldots, n$. Let $Y = \{y \in \mathbb{R}^n_+ : y_1 \leq y_2 \leq \cdots \leq y_n\}$. The set Y is a lattice.

Example 28   *Let N be a finite set ground set and A a finite set of ordered pairs of N. For each $(i, j) \in A$ we have a real number c_{ij}. Let $X = \{x : x_i - x_j \leq c_{ij} \; \forall (i, j) \in A\}$. Assuming X is feasible, then X is a lattice. To see why choose $x, y \in X$. Consider $x \vee y$. Pick an $(i, j) \in A$. Suppose $\max\{x_i, y_i\} = x_i$. Then*

$$\max(x_i, y_i) - \max(x_j, y_j) = x_i - \max(x_j, y_j) \leq x_i - x_j \leq c_{ij}.$$

A similar argument applies to $x \wedge y$.

Definition 7.2   *Let $X \subset \mathbb{R}^n$ be a lattice. An element $x^* \in X$ is the **greatest element** (least) of X if $x^* \geq x$ ($x^* \leq x$) for all $x \in X$.*

Not every lattice has a greatest or least element. The set $[0, \infty)$ is a lattice and has no greatest element. The next theorem gives a sufficient condition for the existence of a greatest or least element.

Theorem 7.3   *If $X \subset \mathbb{R}^n$ is a non-empty compact lattice it has a greatest and least element.*

Proof   We prove that X has a greatest element. A similar proof establishes the existence of a smallest element. For each $i \in \{1, 2, \ldots, n\}$ choose a $z^i \in X$ that maximizes the ith coordinate, i.e., $z^i \in \arg\max_{x \in X} x_i$. Compactness of X ensures that a z^i exists for each i. Let $y = z^1 \vee z^2 \vee z^3 \vee \cdots \vee z^n$. Since X is a lattice,

$y \in X$. Further, $y = (z_1^1, z_2^2, \ldots, z_n^n)$ and by the definition of the z^i's, $y \geq x$ for all $x \in X$.   ∎

Definition 7.4   *Let X be a lattice and $f: X \to \mathbb{R}$. The function f is called* **supermodular** *if for all $z, z' \in X$:*

$$f(z) + f(z') \leq f(z \vee z') + f(z \wedge z').$$

We defer an interpretation of supermodularity till later. Here are some examples of supermodular functions.

1. $f(x_1, x_2) = x_1 x_2$ is supermodular on $\mathbb{R}^2$.
2. $f(x_1, x_2, \ldots, x_n) = x_1^{a_1} x_2^{a_2} \cdots x_n^{a_n}$ is supermodular on $\mathbb{R}_+^n$ when $a_i \geq 0$ for all i.
3. $f(x) = \min_i a_i x_i$ is supermodular on $\mathbb{R}^n$ when $a_i \geq 0$ for all i.

Example 29   *We show that $f(x_1, x_2) = x_1 x_2$ is supermodular on $\mathbb{R}^2$. Choose any two vectors $x = (x_1, x_2)$ and $y = (y_1, y_2)$. Now $f(x) = x_1 x_2$ and $f(y) = y_1 y_2$. Also $f(x \vee y) = \max(x_1, y_1) \max(x_2, y_2)$ and $f(x \wedge y) = \min(x_1, y_1) \min(x_2, y_2)$. If $x \geq y$ then it is easy to see that $f(x \vee y) + f(x \wedge y) \geq f(x) + f(y)$. Now suppose that $x_1 \geq y_1$ but $x_2 \leq y_2$. Then*

$$
\begin{aligned}
f(x \vee y) + f(x \wedge y) - f(x) - f(y) &= x_1 y_2 + x_2 y_1 - x_1 x_2 - y_1 y_2 \\
&= x_1 (y_2 - x_2) - y_1 (y_2 - x_2) \\
&= (x_1 - y_1)(y_2 - x_2) \geq 0.
\end{aligned}
$$

A similar argument applies for the other cases.

The following properties of supermodular functions are easy to prove.

1.   If f is supermodular on the lattice X then af when $a > 0$ is supermodular on X.
2.   If f and g are supermodular on X then $f + g$ is supermodular on X.

One of the most important properties of supermodularity is that it is preserved under maximization.

Theorem 7.5   *Let $X \subset \mathbb{R}^n$, $Y \subset \mathbb{R}^m$ be two lattices and $f: X \times Y \to \mathbb{R}$ supermodular on $X \times Y$. Let $h(y) = \max_{x \in X} f(x, y)$ be well defined for all $y \in Y$. Then h is supermodular on Y.*

Proof Choose $y^1, y^2 \in Y$ and let $x^1, x^2 \in X$ be such that $h(y^i) = f(x^i, y^i)$ for $i = 1, 2$. Then

$$
\begin{aligned}
h(y^1) + h(y^2) &= f(x^1, y^1) + f(x^2, y^2) \\
&\leq f(x^1 \wedge x^2, y^1 \wedge y^2) + f(x^1 \vee x^2, y^1 \vee y^2) \\
&\leq h(y^1 \wedge y^2) + h(y^1 \vee y^2).
\end{aligned}
$$

■

Given a vector $z \in \mathbb{R}^n$ we will write (z_{-ij}, z'_i, z'_j) to mean the vector obtained from z by replacing components i and j with z'_i and z'_j, respectively.

Definition 7.6 *Let $X \subset \mathbb{R}^n$ be a lattice and $f : X \to \mathbb{R}$. The function f satisfies* **increasing differences** *in every pair of components if for all $z \in X$, distinct i and j and $z'_i \geq z_i$, $z'_j \geq z_j$ we have*

$$
f(z_{-ij}, z'_i, z'_j) - f(z_{-ij}, z'_i, z_j) \geq f(z_{-ij}, z_i, z'_j) - f(z_{-ij}, z_i, z_j).
$$

The dot product operation, $f(x, y) = x \cdot y$, satisfies increasing differences on $\mathbb{R}^{2n}$.

Theorem 7.7 *Let $X \subset \mathbb{R}^n$ and $f : X \to \mathbb{R}$. The function f is supermodular iff it satisfies increasing differences on X.*

Proof One direction is easy and is left as an exercise. Here we prove that increasing differences implies supermodularity. Choose any $x, x' \in X$. If $x \leq x'$ or $x \geq x'$ we are done. So, assume not. By rearranging the coordinates there is a k strictly between 0 and n such that

$$
x \wedge x' = (x'_1, \ldots, x'_k, x_{k+1}, \ldots, x_n)
$$

and

$$
x \vee x' = (x_1, \ldots, x_k, x'_{k+1}, \ldots, x'_n).
$$

For any i, j between 0 and n with $i \leq j$ let

$$
x(i, j) = (x_1, \ldots, x_i, x'_{i+1}, \ldots, x'_j, x_{j+1}, \ldots, x_n).
$$

Notice that $x(0, k) = x \wedge x'$, $x(k, n) = x \vee x'$, $x(0, n) = x'$ and $x(k, k) = x$.

From the increasing differences property, we have for $i < k < j$:

$$f[x(i + 1, j + 1)] - f[x(i, j + 1)] \geq f[x(i + 1, j)] - f[x(i, j)].$$

So, for $k \leq j < n$ we have:

$$f[x(k, j + 1)] - f[x(0, j + 1)] = \sum_{i=0}^{k-1} \{f[x(i + 1, j + 1)] - f[x(i, j + 1)]\}.$$

By increasing differences the last term is greater or equal to

$$\sum_{i=0}^{k-1} \{f[x(i + 1, j)] - f[x(i, j)]\} = f[x(k, j)] - f[x(0, j)].$$

To summarize

$$f[x(k, j + 1)] - f[x(0, j + 1)] \geq f[x(k, j)] - f[x(0, j)].$$

Repeated application of this inequality tells us that that left hand side is at most $f[x(k, n)] - f[x(0, n)]$ while the right hand side is at least $f[x(k, k)] - f[x(0, k)]$. Thus left hand side of this inequality achieves its maximum at $j = n - 1$ while the right hand side attains its minimum when $j = k$. Thus

$$f[x(k, n)] - f[x(0, n)] \geq f[x(k, k)] - f[x(0, k)].$$

That is,

$$f(x \wedge x') - f(x) \geq f(x') - f(x \vee x'),$$

which is the supermodularity condition. ∎

If the function f is twice differentiable it is easy to show using the theorem above that f is supermodular iff $\frac{\partial^2 f}{\partial x_i \partial x_j} \geq 0$ for all $i \neq j$.

Supermodularity (or increasing differences) is used to model the notion of **complementarity** in Economics. Computers and monitors are examples of complementarity. Suppose an agent has a utility function $u(c, m)$ where c is the 'quantity' of computers and 'm' the quantity of monitors. The increasing differences condition would imply, e.g., that:

$$u(c + \delta, m + \theta) - u(c + \delta, m) \geq u(c, m + \delta) - u(c, m),$$

where $\delta, \theta > 0$. The left hand side we can interpret as the marginal value of a monitor when the agent has $c + \delta$ units of a computer. The right hand side is the marginal value of the same when the agent has c units of a computer. Thus the marginal value of a monitor increases with the number of computers.

It is sometimes the case that one does not need increasing differences to hold for every pair of components.

Definition 7.8   *Let X and Y be lattices and $f : X \times Y \to \mathbb{R}$. The function f satisfies increasing differences in (x, y) if for all $x, x' \in X$, $y, y' \in Y$ such that $x \geq x'$ and $y \geq y'$ we have*

$$f(x, y) - f(x', y) \geq f(x, y') - f(x', y').$$

If the inequality holds strictly then we say that f satisfies strictly increasing differences.

The next theorem is an important tool for performing comparative statics exercises. The set X below will be a set of actions of strategies while the set Y will be the set of parameters. The theorem tells us how an optimal choice from X changes as we change the choice of parameter from Y.

Theorem 7.9 (Monotone comparative statics)   *Let $X \subset \mathbb{R}^n$ be a compact lattice, $Y \subset \mathbb{R}^m$ a lattice and $f : X \times Y \to \mathbb{R}$ be a continuous function on X for each fixed $y \in Y$. Suppose that f satisfies increasing differences in (x, y) and is supermodular in x for each fixed y.*

1. *For each fixed $y \in Y$, $\arg\max\{f(z, y) : z \in X\}$ is a non-empty compact lattice of $\mathbb{R}^n$ and admits a greatest element $x(y)$.*
2. *$x(y) \geq x(y')$ whenever $y > y'$.*
3. *If f satisfies strictly increasing differences in (x, y), then $x \geq x'$ for any $x \in \arg\max\{f(z, y) : z \in X\}$ and $x' \in \arg\max\{f(z, y') : z \in X\}$ whenever $y \geq y'$.*

Proof   Non-emptiness and compactness of $\arg\max\{f(x, y) : x \in X\}$ for each $y \in Y$ follows from compactness of X and continuity of f. Supermodularity of f implies that $\arg\max\{f(x, y) : x \in X\}$ is a lattice. To see why, suppose not. Choose $x, x' \in \arg\max\{f(z, y) : z \in X\}$ and assume that $x \vee x' \notin \arg\max\{f(z, y) : z \in X\}$. Then

$$f(x \vee x', y) < f(x, y) = f(x', y).$$

Supermodularity of f for each fixed y implies

$$f(x \vee x', y) + f(x \wedge x', y) \geq f(x, y) + f(x', y).$$

Since $f(x \vee x', y) < f(x, y) = f(x', y)$ it follows that that $f(x \wedge x', y) > f(x, y)$, a contradiction. A similar argument applies when we assume that $x \wedge x' \notin \arg\max\{f(z, y) : z \in X\}$. Existence of a largest element follows from Theorem 7.3.

To prove the second part observe that for any $x \in \arg\max\{f(z, y) : z \in X\}$ and $x' \in \arg\max\{f(z, y'): z \in X\}$ we have

$$0 \le f(x', y') - f(x \wedge x', y') \le f(x \vee x', y') - f(x', y')$$
$$\le f(x \vee x', y) - f(x, y) \le 0.$$

The first inequality follows from the choice of x'. The second from supermodularity and the third by increasing differences. We conclude that all inequalities hold as equalities. Now choose $x = x(y)$ and $x' = x(y')$. From the chain of (in)equalities we deduce that $x \vee x' \in \arg\max\{f(z, y): z \in X\}$. But, x is the unique greatest element of $\arg\max\{f(z, y): z \in X\}$ and so $x \ge x \vee x'$, i.e., $x \ge x'$.

The third part is a trivial extension of the second part. ∎

A function on a lattice $X \subset \mathbb{R}^n$ is called **non-decreasing** (also called **isotone**) if for all $x, y \in X$ with $x \le y$ we have $f(x) \le f(y)$. A **non-increasing** (also called **anti-tone**) function is defined similarly.

Theorem 7.10 (Tarski's[1] fixed point theorem) *Let $X \subset \mathbb{R}^n$ be a compact lattice. Let $f: X \to X$ be a non-decreasing function. Then there is an $x^* \in X$ such that $f(x^*) = x^*$.*

Proof Let $X' = \{x \in X: f(x) \ge x\}$. Now $X' \ne \varnothing$. To see why suppose not. Then for all $x \in X$ we have $f(x) < x$. Pick $\bar{x}$ to be the least element of X, such an element exists since X is compact. Then $f(\bar{x}) < \bar{x}$ which is a contradiction since $f(\bar{x}) \in X$.

Consider the set $\{x \in X: x \ge z \forall z \in X'\}$. It is easy to see that this compact and forms a lattice and hence has a least element which we denote $\inf\{x \in X: x \ge z \forall z \in X'\}$.

Let $x^* = \inf\{x \in X: x \ge z: \forall z \in X'\}$. Since X is a compact lattice this is well defined. We show that $x^* \in X'$. Let $y^* = f(x^*)$. Since f is non-decreasing,

$$x^* \le \inf\{x \in X: x \ge f(z): \forall z \in X'\} \le \inf\{x \in X: x \ge f(x^*)\} = y^*.$$

Here $\inf\{x \in X: x \ge f(x^*)\}$ is the least element of the compact lattice $\{x \in X: x \ge f(x^*)\}$.

Since f is non-decreasing, for all $x \in X'$ we have

$$x \le f(x) \le f(x^*) = y^*.$$

Since f is non-decreasing we have $f(x^*) \le f(y^*) = z^*$. As $y^* \ge x^*$ it follows that $z^* \ge y^*$.

Since $f(y^*) = z^* \ge y^*$ it follows that $y^* \in X'$. But $y^* \ge x^*$ and x^* is the greatest element of X'. Thus $f(x^*) = y^* = x^*$, i.e., f has a fixed point. ∎

A careful reading of the proof suggests an algorithm for computing a fixed point that resembles the algorithm used to prove the Banach theorem. Let $\bar{x}$ be the least element of X. Set $x^0 = \bar{x}$ and $x^{i+1} = f(x^i)$. The limit of this sequence is (under appropriate conditions) a fixed point. Notice also that we could have set x^0 to be the largest element of X and produced a sequence (under appropriate conditions) that terminates in a fixed point.

Lemma 7.11 *Let $X \subset \mathbb{R}^n$ be a compact lattice and $f \colon X \to X$ non-decreasing and continuous. If $\bar{x}$ is the least element of X, the sequence $x^{n+1} = f(x^n)$ with $x^0 = \bar{x}$ converges to a fixed point of f.*

Proof Since f is non-decreasing and X compact, the sequence $\{x^n\}_{n \geq 1}$ is non-decreasing and therefore has a limit $x^* \in X$. Since x^{2n} and x^{2n+1} both converge to x^* and $x^{2n+1} = f(x^{2n})$ it follows by the continuity of f that $x^* = f(x^*)$. ∎

The proof makes no special use of the fact that the sequence begins with $\bar{x}$. The role of the least element of X is explained after Corollary 7.16.

7.1 Abstract lattices

Thus far our discussion has focused on lattices with respect to elements of $\mathbb{R}^n$. Lattices are actually more general than this.

A **binary relation**, $\succeq$ on a set X specifies for each pair $x, y \in X$ whether $x \succeq y$ is true or not. If X is some set of males, an example of a binary relation on X would be 'parent of'. Thus $x \succeq y$ if an only if x is the father of y.

Definition 7.12 *A binary relation $\succeq$ on a set X is a **partial order** if it satisfies the following three conditions for all x, y, and $z \in X$*

- **Transitivity**: $x \succeq y$ *and* $y \succeq z$ *imply* $x \succeq z$.
- **Reflexivity**: $x \succeq x$.
- **Antisymmetry**: $x \succeq y$ *and* $y \succeq x$ *imply* $x = y$.

The set of vectors in $\mathbb{R}^n$ with the usual inequality relation is a partial order. More interesting is the set of subsets of a finite set N. If $A, B \subseteq N$, define $A \succeq B$ if $B \subseteq A$. Then $\succeq$ defines a partial order. The binary relation 'parent of' is not a partial order since it violates reflexivity.

The set of real numbers with respect to the inequality relationship is a partial order that differs from the partial order of subsets in an important way. For any two numbers x and y either $x \leq y$ or $y \leq x$, i.e., any two numbers can be ordered. This is not true for subsets. If A and B are subsets of N it is not always true that $A \subseteq B$ or $B \subseteq A$.

If $(X, \succeq)$ is a partial order and $S \subset X$, an **upper bound** for S will be any $x \in X$ such that $x \succeq s$ for all $s \in S$. A **lower bound** is defined similarly. If $x \in S$ is an upper bound (lower bound) for S, then x is called a **greatest element (least**

element) of S. A **maximal element** of S is an element $s \in S$ with no $x \in S$ such that $x \succ s$. Every greatest element is a maximal element but the converse is not true.

As an example, recall from Section 6.4, the set of feasible allocations along with the relationship of Pareto domination. This forms a partial order. A Pareto optimal allocation would be a maximal element. If there were at least two Pareto optimal allocations, there could be no greatest element.

If the set of upper bounds of S has a least element it is called the **least upper bound** of S and denoted $\sup_X(S)$. Similarly, the largest element of the set of lower bounds of S, if it exists, is called the **greatest lower bound** of S and denoted $\inf_X(S)$. The dependence on X in the choice of notation is important. To see why consider $X = \mathbb{R}^1$ and $\bar{X} = [0, 3) \cup \{5\}$. Both are partial orders with respect to the inequality relationship. Let $S = [0, 3)$. Then $\sup_X(S) = 3$ while $\sup_{\bar{X}}(S) = 5$.

Definition 7.13 $(X, \succeq)$ is a **lattice** if $\succeq$ is a partial order on X and every two element subset of X has a least upper bound and greatest lower bound in X:

$$x \vee y = \sup_X\{x, y\} \quad [\textbf{join}],$$

$$x \wedge y = \inf_X\{x, y\} \quad [\textbf{meet}].$$

The partial order of subsets is a lattice with $A \wedge B = A \cap B$ and $A \vee B = A \cup B$.

Definition 7.14   *A lattice* $(X, \succeq)$ *is called* **compact** *if* $\sup_X(S)$ *and* $\inf_X(S)$ *exists for all* $S \subseteq X$.

With these definitions the theorems obtained previously hold even in this more general setting.

Definition 7.15   *Suppose a partially ordered set X is a lattice and $K \subset X$. The set K is a* **sublattice** *of X if* $\sup_X\{x, y\}$ *and* $\inf_X\{x, y\}$ *are in K for all $x, y \in K$.*

Given a lattice X and $K \subset X$ it is possible for K to be a lattice without being a sublattice of X. If K is a lattice this means $\inf_K\{x, y\} \in K$, however $\inf_K\{x, y\}$ need not equal $\inf_X\{x, y\}$. As an example let $X = \mathbb{R}^2$ and $K = \{(0, 0), (2, 1), (1, 2), (3, 3)\}$. The set K is a lattice but is not a sublattice of X because:

$$\sup_K\{(2, 1), (1, 2)\} = (3, 3) \neq (2, 2) = \sup_X\{(2, 1), (1, 2)\}.$$

With this distinction in mind we can state the following corollary of Tarski's theorem.

Corollary 7.16 *Let* $(X, \succeq)$ *be a compact lattice and* $f: X \to X$ *be non-decreasing. The set T of fixed points is a compact lattice with least element* $\sup_X(\{x \in X: f(x) \succeq x\})$ *and greatest element* $\inf_X(\{x \in X: x \succeq f(x)\})$.

The fixed point produced by Lemma 7.11 is the least element of the lattice of fixed points. If the sequence had originated with the greatest element of X, it would have terminated in the greatest fixed point of the lattice of fixed points.

The lattice of fixed points need not be a sublattice of X. As an example let $X = \{(0, 0), (1, 0), (2, 0), (0, 1), (0, 2), (1, 1), (1, 2), (2, 1), (2, 2)\}$. It is easily verified that X is a lattice. Define $f: X \to X$ as

$$f(i, j) = \begin{cases} (i, j), & (i, j) \notin \{(1, 1), (1, 2), (2, 1)\}, \\ (2, 2), & (i, j) \in \{(1, 1), (1, 2), (2, 1)\}. \end{cases}$$

The set T of fixed points is $\{(0, 0), (0, 1), (1, 0), (2, 1), (1, 2)\}$. Notice that $\sup_X\{(1, 0), (0, 1)\} = (1, 1) \notin T$.

The lattice structure and monotonicity allows an analog of Theorem 7.9 for fixed points.

Theorem 7.17 *Let* $(X, \succeq)$ *be a compact lattice and* $(Y, \succeq')$ *a lattice. Let* $f: X \times Y \to X$ *be non-decreasing on* $X \times Y$. *If* $x^*(y)$ *is the least fixed point for each* $y \in Y$, *then* $x^*(y)$ *is non-decreasing in* y.

Proof By Tarski's theorem $x^*(y)$ is well defined for each y. Furthermore, by our proof of Tarski's theorem $x^*(y) = \sup_X\{x: f(x, y) \succeq x\}$. Choose any $y' \succeq' y$. Since f is non-decreasing, $\{x: f(x, y) \succeq x\} \subseteq \{x: f(x, y') \succeq x\}$ from which the result follows. ∎

7.2 Application: supermodular games

An n-person game is called supermodular if the strategy set S^i of each player i is a compact lattice and the payoff function $u^i(s^i, s^{-i})$ for each player is supermodular in $s^i \in S^i$ for each fixed $s^{-i} \in \Pi_{j \neq i} S^j$ and satisfies increasing differences in (s^i, s^{-i}).

Theorem 7.18 *Every n-person supermodular game has a Nash equilibrium.*

Proof For each player i and $s \in \Pi_{j=1}^n S^j$ let

$$B^i(s^{-i}) = \arg\max\{u^i(t, s^{-i}): t \in S^i\}.$$

From Theorem 7.9 it follows that $B^i(s)$ is a lattice and has a greatest element $b_*^i(s^{-i})$. From the same Theorem it follows that $b_*^i(s)$ is a non-decreasing function and so is

$$b_*(s) = (b_*^1(s^{-1}), \dots, b_*^n(s^{-n})).$$

Since each S^i is a compact lattice so is $\Pi_i S^i$ and $b_*: \Pi_i S^i \rightarrow \Pi_i S^i$ is non-decreasing. Existence of equilibrium now follows from Tarski's theorem.   ■

Given a competitive situation modeled by a game one would like to be able to do 'comparative statics' on the game. For example, if the firms costs change how does that effect the equilibrium price? This is difficult to do when the game has multiple equilibria. Which equilibrium does one pick out in making the before and after comparison? Supermodular games have the property that their Nash equilibria form a lattice. There is thus a natural way to (partially) order the equilibria of a game. One can also make comparisons of equilibria after parameter changes by looking at the maximal or minimal equilibria of the lattice of equilibrium outcomes.

7.3   Application: transportation problem

Procrustes & Sons[2] manufactures soma at a finite set of locations called S (supply nodes). The maximum amount that a node $i \in S$ can supply is s_i. Buyers are located at a finite number of locations called D (demand nodes). The total amount demanded by a buyer $j \in D$ is d_j. The cost per unit incurred to ship soma from supply node i to demand node j is c_{ij}. The firm must meet the demand of each buyer and do so at minimum cost. The problem faced by the firm can be formulated as a linear program.[3] To ensure feasibility we assume that $\sum_{i \in S} s_i \geq \sum_{i \in D} d_i$, i.e., supply exceeds demand.

Let x_{ij} denote the amount of soma shipped from supply node i to demand node j. Since no more can be supplied from supply node i than is available we must have

$$\sum_{j \in D} x_{ij} \leq s_i, \quad \forall i \in S.$$

The amount shipped to demand node j must be at least as large the demand at node j, i.e.,

$$\sum_{i \in S} x_{ij} \geq d_j, \quad \forall j \in D.$$

We could enforce equality here but is unnecessary since, it will follow automatically from trying to minimize shipping costs.

Total shipping costs will be $\sum_{i \in S} \sum_{j \in D} c_{ij} x_{ij}$. The problem facing Procrustes & Co. is

$$\min \sum_{i \in S} \sum_{j \in D} c_{ij} x_{ij}$$

$$\text{s.t.} \quad -\sum_{j \in D} x_{ij} \geq -s_i, \quad \forall i \in S,$$

$$\sum_{i \in S} x_{ij} \geq d_j, \quad \forall j \in D,$$

$$x_{ij} \geq 0, \quad \forall i \in S, \ j \in D.$$

If $s \in \mathbb{R}_+^{|S|}$ and $d \in \mathbb{R}_+^{|D|}$ are the vectors of supply and demand respectively, denote the optimal value of the objective function by $c(-s, d)$.

Let p_i denote dual variable associated with the ith supply constraint and q_j the dual variable associated with the jth demand constraint. The dual program is

$$c(-s, d) = \max -\sum_{i \in S} s_i p_i + \sum_{j \in D} d_j q_j$$

$$\text{s.t.} \quad -p_i + q_j \geq c_{ij}, \quad \forall i \in S, \ j \in D,$$

$$p_i, q_j \geq 0, \quad \forall i \in S, \ j \in D.$$

Lemma 7.19 *The set of feasible solutions to the dual problem is a lattice.*

Proof Pick two dual feasible solutions $(-p, q)$ and $(-p', q')$. Consider $(-p, q) \vee (-p', q')$. The ith component is $\max(-p_i, -p'_i)$ and the jth component will be $\max(q_j, q'_j)$. We must show that

$$\max(-p_i, -p'_i) + \max(q_j, q'_j) \geq c_{ij}.$$

Without loss of generality suppose that $q_j \geq q'_j$. Then

$$\max(-p_i, -p'_i) + \max(q_j, q'_j) = \max(-p_i, -p'_i) + q_j$$

$$\geq \max(-p_i, -p'_i) + c_{ij} - p_i \geq c_{ij}. \quad \blacksquare$$

Theorem 7.20 $c(-s, d)$ *is supermodular in* $(-s, d)$.

Proof The objective function of the dual problem is the sum of two dot products and so is supermodular in $(-s, d, p, q)$. The feasible region is a lattice. For each choice of $(-s, d)$ an optimal solution to the dual exists. The theorem now follows from Theorem 7.5. $\quad \blacksquare$

Part (1) of Theorem 7.9 implies that the set of optimal dual solutions forms a lattice. The set of optimal dual solutions is not compact, but is bounded below by zero. A simple modification of the proof of Theorem 7.3 yields the existence of a smallest element. Part (2) of Theorem 7.9 implies that the set of optimal dual solutions is increasing in $(-s, d)$. Thus, if s_i were to decrease (i.e., $-s_i$ increases), we expect the optimal value of p_i to go up. If d_j increases, we expect the optimal value of q_j to increase.

7.4   Application: efficient assignment and the core

We revisit the problem of finding the efficient assignment discussed in Section 4.9. As before M is a set of distinct indivisible goods and N the set of agents. We denote by $V(N)$ the total value of an efficient assignment. Thus,

$$V(N) = \max \sum_{j \in N} \sum_{i \in M} v_{ij} x_{ij}$$

$$\text{s.t.} \sum_{j \in N} x_{ij} \leq 1, \quad \forall i \in M,$$

$$\sum_{i \in M} x_{ij} \leq 1, \quad \forall j \in N,$$

$$0 \leq x_{ij} \leq 1, \quad \forall i \in M, \ j \in N.$$

Before continuing, the reader is urged to review Section 4.9 in particular the portion about supporting prices.

We will also be interested in the value of an efficient assignment when we restrict attention to a subset S of agents. We call this problem P(S).

$$V(S) = \max \sum_{j \in N} \sum_{i \in M} v_{ij} x_{ij}$$

$$\text{s.t.} \sum_{j \in N} x_{ij} \leq 1, \quad \forall i \in M,$$

$$\sum_{i \in M} x_{ij} \leq 1, \quad \forall j \in S,$$

$$\sum_{i \in M} x_{ij} \leq 0, \quad \forall j \notin S,$$

$$0 \leq x_{ij} \leq 1, \quad \forall i \in M, \ j \in S.$$

Note that the constraint matrix is still totally unimodular. The dual to problem P(S), denoted DP(S) is

$$\min \sum_{i \in M} p_i + \sum_{j \in S} \lambda_j$$

$$\text{s.t.} \ p_i + \lambda_j \geq v_{ij}, \quad \forall i \in M, \ j \in N.$$

Theorem 7.21 *V(·) is non-decreasing and submodular.*

Proof Let $d \in \mathbb{R}^{|N|}$ be a 0–1 vector and let $B^{|N|}$ be the set of all 0–1 vectors in $\mathbb{R}^{|N|}$. Let

$$f(d) = \max \sum_{j \in N} \sum_{i \in M} v_{ij} x_{ij}$$

$$\text{s.t.} \quad \sum_{j \in N} x_{ij} \leq 1, \quad \forall i \in M,$$

$$\sum_{i \in M} x_{ij} \leq d_j, \quad \forall j \in N,$$

$$0 \leq x_{ij} \leq 1, \quad \forall i \in M, \ j \in N.$$

By the duality theorem

$$f(d) = \min \sum_{i \in M} p_i + \sum_{j \in N} d_j \lambda_j$$

$$\text{s.t.} \quad p_i + \lambda_j \geq v_{ij}, \quad \forall i \in M, \ j \in N,$$

$$p_i, \lambda_j \geq 0, \quad \forall i \in M, \ j \in N.$$

We make a change of variables: $w_j = -\lambda_j$ for all $j \in N$. With this change

$$f(d) = \min \sum_{i \in M} p_i - \sum_{j \in N} d_j w_j$$

$$\text{s.t.} \quad p_i - w_j \geq v_{ij}, \quad \forall i \in M, \ j \in N,$$

$$p_i \geq 0, \quad w_j \leq 0, \quad \forall i \in M, \ j \in N.$$

The set of feasible dual solutions forms a lattice with respect to the partial order $(p, w) \succeq (p', w')$ if and only if $(p, w) \geq (p, w')$.[4] The objective function of this last program is submodular. From Theorem 7.5 we deduce that $f(d)$ is submodular on the lattice $B^{|N|}$. If we set $d_j = 1$ for all $j \in S$ and zero otherwise, it follows that $V(S) = f(d)$. ∎

We can associate with the problem of finding an efficient assignment a cooperative game. To define it we introduce a new agent (not in N) called the seller, s. The seller is assumed to own all the goods. The characteristic function u is defined as

1. $u(S) = 0, \forall S \subseteq N,$
2. $u(S \cup s) = V(S), \forall S \subseteq N.$

An interpretation is that for a coalition of agents S to generate value, they must include the seller. The core, $C(u, N \cup \{s\})$, of this game is:

$$\sum_{j \in N} \mu_j + \mu_s = u(N \cup s) = V(N),$$

$$\sum_{j \in \{S \cup s\}} \mu_j + \mu_s \geq u(S \cup s) = V(S),$$

$$\sum_{j \in S} \mu_j \geq u(S) = 0, \quad \forall S \subset N.$$

Recall that an outcome in the core is a division of value that no coalition of agents can 'block'.

Lemma 7.22 $C(u, N \cup \{s\})$ *is non-empty.*

Proof   Set $\mu_j = 0$ for all $j \in N$ and $\mu_s = V(N)$. For any $S \subseteq N$ we have $\sum_{i \in S} \mu_i = 0 = u(S)$. For any $S \cup \{s\}$ we have that $\sum_{i \in S \cup \{s\}} \mu_i = V(N) \geq V(S)$ since $V(\cdot)$ is non-decreasing (see Lemma 7.23).   ∎

Let (λ^*, p^*) be an optimal dual solution to $D(N)$. Recall that we interpret p^* as a price vector and λ^* as a vector whose jth component gives the surplus of agent j at prices p^*.

Lemma 7.23   *The efficient assignment produces an outcome that is in $C(u, N \cup \{s\})$ in the sense that $\mu_j = \lambda_j^*$ and $\mu_s = \sum_{i \in M} p_i^*$ is a point in $C(u, N \cup \{s\})$.*

Proof   By the definition of the dual

$$u(N \cup \{s\}) = V(N) = \sum_{j \in N} \lambda_j^* + \sum_{i \in M} p_i^* = \sum_{j \in N \cup \{s\}} \mu_j.$$

Furthermore (λ^*, p^*) is a feasible solution to $DP(S)$ for all $S \subseteq N$. By dual feasibility we have that

$$u(S \cup \{s\}) = V(S) \leq \sum_{j \in S} \lambda_j^* + \sum_{i \in M} p_i^* = \sum_{j \in S \cup \{s\}} \mu_j.$$ ∎

The quantity $V(N) - V(N \setminus j)$ for all $j \in N$ is called the **marginal product** of agent j. It represents agent j's 'added value' to the coalition $N \cup \{s\}$. For any outcome in the core, no agent can obtain a surplus that exceeds its marginal product. To see why, consider the following two restraints from the definition of

$C(u, N \cup \{s\})$:

$$\sum_{j \in N} \mu_j + \mu_s = V(N),$$

$$\sum_{j \in N \setminus k} \mu_j + \mu_s \geq V(N \setminus k).$$

Negating the second and adding to the first yields $\mu_k \leq V(N) - V(N \setminus k)$. Remarkably there is a point in the core that gives to each agent in N their marginal product.

Theorem 7.24 *There is a point $\mu \in C(u, N \cup \{s\})$ such that $\mu_j = V(N) - V(N \setminus j)$ for all $j \in N$.*

Proof Set $\mu_j = V(N) - V(N \setminus j)$ for all $j \in N$ and $\mu_s = V(N) - \sum_{j \in N}[V(N) - V(N \setminus j)]$. With this choice we have $\sum_{j \in N} \mu_j + \mu_s = u(N \cup \{s\})$. To complete the proof we show that $\sum_{j \in S} \mu_j + \mu_s \geq u(S \cup \{s\})$ for all $S \subset N$.

Observe that

$$\sum_{j \in S} \mu_j + \mu_s - V(S) = V(N) - V(S) - \sum_{j \in N \setminus S}[V(N) - V(N \setminus j)].$$

We use the submodularity of $V(\cdot)$ to show that the right-hand side of the above is non-negative. Let $N \setminus S = \{j_1, j_2, \ldots, j_k\}$ and take j_0 to be the empty set. Then, from increasing differences,

$$V(S \cup \{j_1, \ldots, j_r\}) - V(S \cup \{j_1, \ldots, j_{r-1}\}) \geq V(N) - V(N \setminus j_r).$$

Therefore,

$$V(N) - V(S) = \sum_{r=1}^{k}[V(S \cup \{j_1, \ldots, j_r\}) - V(S \cup \{j_1, \ldots, j_{r-1}\})]$$

$$\geq \sum_{j \in N \setminus S}[V(N) - V(N \setminus j)].$$

Last, $\sum_{j \in S} \mu_j \geq 0$ follows from the non-negativity of marginal products which in turn follows from $V(\cdot)$ being non-decreasing. ∎

We now establish a converse to Lemma 7.23. We prove that every point in the core corresponds to an optimal solution to D(N).

Lemma 7.25 *For every $\mu \in C(u, N \cup \{s\})$ there is a vector $p \in \mathbb{R}^{|M|}$ such that (λ, p) is an optimal solution to D(N) where $\lambda_j = \mu_j$ for all $j \in N$.*

Proof   Given $\mu \in C(u, N \cup \{s\})$ set $\lambda_j = \mu_j$ for all $j \in N$. We show that there is a $p \in \mathbb{R}^{|M|}$ such that $\sum_{i \in M} p_i = \mu_s$ and $\lambda_j + p_i \geq v_{ij}$ for all $i \in M$ and $j \in N$.

For each i set $p_i = \max_{j \in N}(v_{ij} - \lambda_j)$ for all $i \in M$ and let $N_i = \arg\max_{j \in N}(v_{ij} - \lambda_j)$. If

$$\sum_{j \in N} \lambda_j + \sum_{i \in M} p_i \leq V(N)$$

we are done. Simply raise the value of one of the p_i's until equality is reached. So, suppose not. In this case:

$$\sum_{j \in N} \lambda_j + \sum_{i \in M} p_i > V(N).$$

We now assign each $i \in M$ to at most one $j \in N_i$ so that no $j \in N$ is assigned more than one good from M. Amongst all such assignments, choose one that maximizes the number of agents in N who receive a good. Let S be that set of agents, and G the set of goods assigned to agents in S. Thus $|S| = |G|$ and each agent in S is assigned exactly one good in G and each good in G is assigned to exactly one agent in S. Goods in $M \backslash G$ are not assigned and agents in $N \backslash S$ receive no goods. Notice, if good i is assigned to agent j then $j \in N_i$. We show that (λ, p) is an optimal solution to D(S).

Let x^* be a feasible solution to P(S) defined by setting $x_{ij} = 1$ if good $i \in G$ is assigned to agent $j \in S$ and zero in all other cases. Now we verify the complementary slackness conditions to establish optimality. Specifically we must prove that $x_{ij}^*(\lambda_j + p_i - v_{ij}) = 0$ for all $i \in M$ and $j \in S$. If $x_{ij}^* = 0$ this clearly true. When $x_{ij}^* = 1$, then $j \in N_i$, i.e., $p_i = v_{ij} - \lambda_j$ and so complementary slackness holds.

Hence

$$V(S) = \sum_{i \in M} p_i + \sum_{j \in S} \lambda_j > V(N) - \sum_{j \in N \backslash S} \lambda_j$$

$$\Rightarrow V(N) - V(S) < \sum_{j \in N \backslash S} \lambda_j \leq \sum_{j \in N \backslash S} [V(N) - V(N \backslash j)]$$

which violates submodularity of $V(\cdot)$. The last inequality follows from the observation preceeding Theorem 7.24.   ∎

Hence amongst the optimal dual solutions to P(N) is one that gives to each agent their marginal product. Equivalently, amongst all prices that support an efficient allocation, there is one that leaves each agent with a surplus equal to their marginal product.

Using the marginal value theorem of linear programming, we can identify which optimal dual solution gives each agent their marginal product. Amongst all optimal solutions to D(N) select the one that minimizes $\sum_{i \in N} p_i$. Given the lattice structure used in Theorem 7.21, this is the least element of the lattice of feasible dual solutions.

7.5 Application: stable matchings

Let M be a set of men and W a set of women and suppose that $|M| = |W|$. Each $m \in M$ has a strict preference ordering over the elements of W and each $w \in W$ has a strict preference ordering over the men. The preference ordering of an individual i will be denoted $>^i$ and $x >^i y$ will mean that agent i ranks x above y.

A **matching** is an assignment of men to women such that each man is assigned to one woman and vice-versa. A matching is called **unstable** if there are two men m, m' and two women w, w' such that

1. m is matched to w,
2. m' is matched to w',
3. and $w' >^m w$ and $m >^{w'} m'$.

The pair (m, w') is called a **blocking pair**. A matching that has no blocking pairs is called **stable**.

Example 30 *Men occupy the rows and women the columns. The first entry in each cell is the rank that the man corresponding to that row assigns to the woman corresponding to the relevant column. The second entry is the rank that the woman corresponding to that column assigns to the man in the corresponding row:*

$M-W$	w_1	w_2	w_3
m_1	$(2, 1)$	$(1, 2)$	$(3, 1)$
m_2	$(1, 3)$	$(3, 3)$	$(2, 3)$
m_3	$(1, 2)$	$(2, 1)$	$(3, 2)$

Consider the matching $\{(m_1, w_1), (m_2, w_2), (m_3, w_3)\}$. This is an unstable matching since (m_1, w_2) is a blocking pair. The matching $\{(m_1, w_1), (m_3, w_2), (m_2, w_3)\}$ is stable.

Given the preferences of the men and women, is it always possible to find a stable matching? Remarkably, yes. This was first established by David Gale and Lloyd Shapley[5] using what is now called the deferred proposal algorithm. Here we give a proof using Tarski's fixed point theorem.[6]

Call an assignment of women to men such that each man is assigned to at most one woman but a woman may be assigned to more than one man a **male semi-matching**. Call the analogous object for women a **female semi-matching**. For example, assigning each man his first choice would be a male semi-matching. Assigning each woman her third choice would be an example of a female semi-matching.

A pair of male and female semi-matchings will be called a **semi-matching** which we will denote by μ, ν etc. An example of a semi-matching would consist of each man being assigned his first choice and each woman being assigned her last choice.

The woman assigned to the man m under the semi-matching μ will be denoted $\mu(m)$. If man m is assigned to no woman under μ, then $\mu(m) = m$. Similarly for $\mu(w)$. Next we define a partial order over the set of semi-matchings. Write $\mu \succeq \nu$ if

1. $\mu(m) >^m \nu(m)$ or $\mu(m) = \mu(m)$ for all $m \in M$, and,
2. $\mu(w) <^w \nu(w)$ or $\mu(w) = \nu(w)$ for all $w \in W$.

Roughly speaking, $\mu \succeq \nu$ if all the men are better off under μ than in ν and all the women are worse off under μ than in ν.

Next we define the meet and join operations. Given two semi-matchings μ and ν define $\lambda = \mu \vee \nu$ as follows:

1. $\lambda(m) = \mu(m)$ if $\mu(m) >^m \nu(m)$ otherwise $\lambda(m) = \nu(m)$,
2. $\lambda(w) = \mu(w)$ if $\mu(w) <^w \nu(w)$ otherwise $\lambda(w) = \nu(w)$.

Define $\lambda' = \mu \wedge \nu$ as follows:

1. $\lambda'(m) = \mu(m)$ if $\mu(m) <^m \nu(m)$ otherwise $\lambda(m) = \nu(m)$,
2. $\lambda(w) = \mu(w)$ if $\mu(w) >^w \nu(w)$ otherwise $\lambda(w) = \nu(w)$.

With these definitions it is easy to check that the set of semi-matchings form a compact lattice.

Now define a function f on the set of semi-matchings that is non-decreasing. Given a semi-matching μ define $f(\mu)$ to be the following semi-matching:

1. $f(\mu)(m)$ is man m's most preferred woman from the set $\{w: m >^w \mu(w), m = \mu(w)\}$. If this set is empty set $f(\mu)(m) = m$.
2. $f(\mu)(w)$ is woman w's most preferred man from the set $\{m: w >^m \mu(m), w = \mu(m)\}$. If this set is empty set $f(\mu)(w) = w$.

It is clear that f maps semi-matchings into semi-matchings.

Theorem 7.26   *There is a semi-matching μ such that $f(\mu) = \mu$. Furthermore μ is a stable matching.*

Proof   We use Tarski's theorem. It suffices to check that f is non-decreasing. Suppose $\mu \succeq \nu$. Pick any $m \in M$. From the definition of $\succeq$, the women are worse off under μ than in ν. Thus

$$\{w: m >^w \nu(w)\} \subseteq \{w: m >^w \mu(w)\}$$

and so $f(\mu)(m) >^m f(\nu)(m)$ or $f(\mu)(m) = f(\nu)(m)$. A similar argument applies for each $w \in W$. Thus f is non-decreasing.

Since the conditions of Tarski's theorem hold, it follows that there is a semi-matching μ such that $f(\mu) = \mu$. We show that the semi-matching is a stable matching.

By the definition of a semi-matching we have for every $m \in M$, $\mu(m)$ single valued as is $\mu(w)$ for all $w \in W$. To show that μ is matching, suppose not. Then there is a pair $m_1, m_2 \in M$, say, such that $\mu(m_1) = \mu(m_2) = w^*$. Since $f(\mu) = \mu$ it follows that w^* is m_1's top ranked choice in $\{w: m_1 >^w \mu(w), m_1 = \mu(w)\}$ and m_2's top ranked choice in $\{w: m_2 >^w \mu(w), m_2 = \mu(w)\}$. From this we deduce that $\mu(w^*) = m_3$ where $m_1, m_2 >^{w^*} m_3$. However, $m_3 = \mu(w^*) = f(\mu^*)(w^*)$ which is woman w^*'s top ranked choice in $\{m: w^* >^m \mu(m), \mu(m) = w^*\}$. Since m_1, m_2 are members of this set, we get a contradiction.

To show that the matching μ is stable suppose not. Then there must be a blocking pair (m^*, w^*). Let $w' = \mu(m^*)$ and $m' = \mu(w^*)$, $m' \neq m^*$ and $w^* \neq w'$. Since (m^*, w^*) is blocking, $m^* >^{w^*} m'$ and $w^* >^{m^*} w'$. Now $w' = \mu(m^*) = f(\mu)(m^*)$ which is man m^*'s top ranked choice from $\{w: m^* >^w \mu(w), m^* = \mu(w)\}$. But this set contains w^* which is ranked higher by man m^* than w', a contradiction. ∎

Problems

7.1 Show that the set $S = \{(x, y): x - y = 1\}$ is a lattice in $\mathbb{R}^2$.

7.2 Let $S = [0, 1]$ and in each of the following $f(x, y)$ is a function from $S \times \mathbb{R}_+$ into $\mathbb{R}$; $x \in S$ and $y \in \mathbb{R}_+$. Decide which of them are supermodular in (x, y).

    1. $f(x, y) = xy - x^2 y^2$
    2. $f(x, y) = xy - x^2$
    3. $f(x, y) = x/(1 + y)$
    4. $f(x, y) = x(1 + y)$
    5. $f(x, y) = x(y - x)$

7.3 Show that the product of non-negative, non-decreasing supermodular functions on a lattice $X \subseteq \mathbb{R}^n$ is supermodular.

7.4 Let $\{f_k\}_{k\geq 1}$ be a sequence of supermodular functions defined on a lattice $X \subseteq \mathbb{R}^n$. Suppose that $\lim_{k\to\infty} f_k(x) = f(x)$ for all $x \in X$. Show that f is supermodular.

7.5 Let $X \subset \mathbb{R}^n$ be a compact lattice. Show that there is a supermodular function on $\mathbb{R}^n$ such that $X = \arg\max_{x \in \mathbb{R}^n} f(x)$.
Hint: Consider $g(x, z) = -\sum_{i=1}^{n} |x_i - z_i|$.

7.6 Prove Corollary 7.16.

7.7 Consider n firms making imperfectly substitutable products. Firm i has a constant marginal costs of c_i, announces a price p_i and the demand it faces as a function of all prices is

$$d_i(p_1, \ldots, p_n) = a_i - b_i p_i + \sum_{i \neq j} g_{ij} p_j.$$

Here b_i and g_{ij} are strictly positive for all i, j. Firm i's profit as a function of everyone else's price (abbreviated to p^{-i}) and its own price p is $\Pi(p^{-i}, p) = (p - c_i)d_i(p^{-i}, p)$. For each fixed p^{-i} show that $\Pi(p^{-i}, p)$ is supermodular in p.

7.8 Consider two firms making identical products. Firm i has a constant marginal costs of $c_i < 1$, announces a quantity q_i. The price per unit of the good in the market is given by $1 - q_1 - q_2$. The profit of firm i will be $[(1 - q_1 - q_2)q_i - cq_i]$. Consider a game where the firms choose their quantities simultaneously. By treating one firms strategy as being an element of $[0, 1]$ and the other as $[-1, 0]$ show that the game is supermodular.

7.9 Consider a two person game where each players strategy set is $[0, 1]$ and denote by x_i the strategy of player i. The payoff function for player 1 is $u_1(x_1, x_2) = -x_1^2 + x_2$ and for player 2 is $u_2(x_1, x_2) = -x_2^2 + x_1$. Show that this game is supermodular and has a unique equilibrium.

Notes

1 Born Alfred Teitlebaum in 1902, died Alfred Tarski in 1983. In between a road to Damascus conversion to Catholicism. One of the four greatest logicians of all time. In spite of all temptations to belong to other nations, remained a Pole.
2 On the road to Attica the unsuspecting traveller who accepted Procrustes' invitation to lie in his iron bed, would be stretched or shortened, with fatal consequences, to fit the length of the bed. Procrustes was slain by Theseus of Minotaur fame.
3 This model generalizes the one of Section 4.9.
4 Equivalently, $p \geq p'$ and $\lambda \leq \lambda'$.
5 Gale and Shapley (1962).
6 The approach has its roots in Subramanian (1994). It is fleshed out in Adachi (2000).

References

Adachi, H.: 2000, On a characterization of stable matchings, *Economics Letters* **68**(1), 43–9.
Gale, D. and Shapley, L. S.: 1962, College admissions and the stability of marriage, *American Mathematical Monthly* **69**(1), 9–15.
Subramanian, A.: 1994, A new approach to stable matching problems, *SIAM Journal on Computing* **23**(4), 671–701.
Topkis, D. M.: 1998, *Supermodularity and complementarity*, Frontiers of economic research, Princeton University Press, Princeton, N.J.

8 Matroids

Matroids, defined below, are an abstraction of the idea of independence from linear algebra. There are close connections between matroids, optimization and submodularity which will be described here.

8.1 Introduction

Let E be a finite ground set and $\mathcal{I}$ a family of subsets of E.

Definition 8.1 $(E, \mathcal{I})$ is called an **independence system** if

1. $\varnothing \in \mathcal{I}$,
2. $A \subset B \in \mathcal{I} \Rightarrow A \in \mathcal{I}$.

Here are five examples:

1. $E = \{1, 2, 3\}$, $\mathcal{I} = \{\varnothing, (1), (2), (3), (1, 3)\}$.
2. Let A be an $m \times n$ matrix and E the index set of columns of A. Define $\mathcal{I}$ to be the subsets of E that correspond to linearly independent columns of A.
3. Given m disjoint sets E_i, $i = 1, \ldots, m$, let $E = \cup_{i=1}^{m} E_i$. Set $\mathcal{I} = \{F \subseteq E : |F \cap E_i| \leq 1 \ \forall i = 1, \ldots, m\}$.
4. Let $G = (V, E)$ be a graph and $\mathcal{I} = \{F \subseteq E : (V, F) \text{ is acyclic}\}$.
5. Let E be a finite set and t some positive integer that is at most $|E|$. Let $\mathcal{I} = \{S \subseteq E : |S| \leq t\}$.

Elements of $\mathcal{I}$ are called **independent sets**. Non-elements of $\mathcal{I}$ are called **dependent sets**.

Definition 8.2 Let $(E, \mathcal{I})$ be an independence system. For any $T \subseteq E$ a set $B \subseteq T$ is called a **basis** of T or **maximal in** T if $B \in \mathcal{I}$ and $B \cup j \notin \mathcal{I}$ for all $j \in T \setminus B$.

Definition 8.3 If $(E, \mathcal{I})$ is an independence system, a set $C \subseteq E$ is called a **circuit** if it is minimally dependent. That is $C \notin \mathcal{I}$ but $C \setminus j \in \mathcal{I}$ for all $j \in C$.

Example 31   *Let E be the set of columns of the matrix below:*

$$\begin{bmatrix} 2 & 1 & 1 & 1 \\ 1 & -1 & -1 & 1 \end{bmatrix}.$$

A set of columns will be independent if the corresponding columns are linearly independent. The first and third columns form a basis for E. The first, third and fourth columns form a circuit.

Definition 8.4   *The independence system $(E, \mathcal{I})$ is called a **matroid** if for all $T \subseteq E$, all basis of T have the same size. That is, if B and B' are each a basis for T then $|B| = |B'|$.*

In the list of independent systems given previously, only the last four are matroids. The second is called a **matric** matroid, the third is called a **partition** matroid, the fourth is called the **forest** matroid and the fifth is called the **uniform** matroid. The reader should be able to verify that these independence systems are all matroids.

We derive some properties of matroids that will be useful later. In reading the theorems and proofs the reader will find it helpful to keep either the matric or graphic matroid in mind.

Lemma 8.5   *Let $(E, \mathcal{I})$ be an independence system. For any $T \subseteq E$ and independent set $S \subseteq T$ there is a basis B of T such that $S \subseteq B$.*

Proof   If there is a $j \in T \setminus S$ such that $S \cup j \in \mathcal{I}$ then add it to S. Continue. We must eventually enlarge S into a set B such that $B \cup k \notin \mathcal{I}$ for all $k \in T \setminus B$.   ∎

Lemma 8.6   *Let $(E, \mathcal{I})$ be an independence system. Every $T \notin \mathcal{I}$ contains at least one circuit.*

Proof   If T is a circuit we are done. If not there is a $j \in T$ such that $T \setminus j \in \mathcal{I}$. Delete j from T. Repeat. Eventually we obtain a set $C \subseteq T$ such that $C \setminus k \in \mathcal{I}$ for all $k \in C$.   ∎

Theorem 8.7   *Let $(E, \mathcal{I})$ be a matroid and A, B $\in \mathcal{I}$ with $|A| < |B|$. Then there exists $j \in B \setminus A$ such that $A \cup j \in \mathcal{I}$.*

Proof   Suppose not. Then A is a basis for $A \cup B$. Since $B \subset A \cup B$ there is another basis for $A \cup B$ with a cardinality of at least $|B| > |A|$. It can be obtained by adding elements of $(A \cup B) \setminus B$ to B. Thus we get two basis for the same set of different sizes, a contradiction.   ∎

Theorem 8.8   *Let $(E, \mathcal{I})$ be a matroid. For any $A \in \mathcal{I}$ and any $j \in E$, $A \cup j$ contains at most one circuit.*

Proof Suppose not. Choose the smallest A that contradicts the theorem. Then $A \cup j$ contains two circuits C and C'. By the choice of A, $A = (C \cup C') \setminus j$. Choose $a \in C \setminus C'$ and $b \in C' \setminus C$. Let $S = (C \cup C') \setminus \{a, b\}$. If $S \notin \mathcal{I}$, then S contains a circuit D. This implies $(A \setminus a) \cup j$ contains two circuits, D and C'. Since $A \setminus a \in \mathcal{I}$ and is of cardinality one less than that of A, we have a contradiction. So, $S \in \mathcal{I}$. But S is a basis of $C \cup C'$ as is A and $|S| \neq |A|$ a contradiction. ∎

Theorem 8.9 *A set $\mathcal{C}$ of subsets of E are the circuits of a matroid if and only if*

1. $\varnothing \notin \mathcal{C}$.
2. *If $C, C' \in \mathcal{C}$ and $C \subseteq C'$ then $C = C'$.*
3. *If $C, C' \in \mathcal{C}$, $C \neq C'$ and $e \in C \cap C'$ then there exists $D \in \mathcal{C}$ such that $D \subseteq (C \cup C') \setminus e$.*

Proof Suppose first that $\mathcal{C}$ is a collection of circuits for the matroid $(E, \mathcal{I})$. Then the first two statements are obvious. If the third statement is violated for some $C, C' \in \mathcal{C}$ then $B = (C \cup C') \setminus e \in \mathcal{I}$. Then $B \cup e$ contains two circuits which violates Theorem 8.8

Now suppose $\mathcal{C}$ is a collection of sets that satisfies (1–3) of the theorem. Let $\mathcal{I} = \{S: S$ contains no member of $\mathcal{C}\}$. We show that $(E, \mathcal{I})$ is a matroid. In view of (1) and (2) $(E, \mathcal{I})$ is clearly an independence system.

Pick $A \subseteq E$ and let B and B' be two different basis for A. Suppose $|B| < |B'|$. Choose them so that $|B \cap B'|$ is maximized. Pick $e \in B' \setminus B$. Then $B' \cup e$ contains at least one circuit, C. We will use (3) to argue that the circuit is unique.

Suppose not and let C' be the other circuit. By (3) $(C \cup C') \setminus e$ contains a circuit. However, $(C \cup C') \setminus e \subset B$, contradicting the independence of B.

Since $C \not\subseteq B$ there exists $q \in C \setminus B$. Then

$$T = (B \cup e) \setminus q \in \mathcal{I}$$

since C is the only member of $\mathcal{C}$ in $B \cup e$. But $|T \cap B| > |B' \cap B|$, contradicting the choice of B and B'. ∎

8.2   Matroid optimization

We associate with the matroid $(E, \mathcal{I})$ a weight vector w that assigns to each $e \in E$ a weight w_e. The matroid optimization problem is to find an independent set of largest total weight: $\max_{S \in \mathcal{I}} \sum_{e \in S} w_e$. Subsequently we show how to formulate this problem as a linear program. For now we give a direct method for solving the matroid optimization problem.

Greedy algorithm

1.   Order elements of E: $w_1 \geq w_2 \cdots \geq w_n$.
2.   Set $S^0 = \varnothing$ and $t = 1$.
3.   If $w_t \leq 0$, stop and output S^{t-1}.
4.   If $w_t > 0$ and $S^{t-1} \cup t \in \mathcal{I}$ set $S^t = S^{t-1} \cup t$.

5.   If $w_t > 0$ and $S^{t-1} \cup t \notin \mathcal{I}$ set $S^t = S^{t-1}$.
6.   If $t = n$ stop. If $t < n$, set $t = t + 1$ and goto (3).

Example 32   *Let $(E, \mathcal{I})$ be the matric matroid associated with the matrix below.*

$$\begin{bmatrix} 2 & 1 & 1 & 1 \\ 1 & -1 & -1 & 1 \\ 0 & 0 & 0 & 1 \end{bmatrix}.$$

The first column will have weight 7, the second weight 5, the third weight 4, the fourth weight 3. The greedy algorithm will select the first column followed by the second column. Then it will skip the third column (since the first three columns are linearly dependent) and select the fourth column.

The greedy algorithm computes a maximum weight basis for a matroid.

Theorem 8.10   *Let $(E, \mathcal{I})$ be a matroid. For every weight vector w, the greedy algorithm finds a maximum weight independent set.*

Proof   Let $G = \{e_1, e_2, \ldots, e_m\}$ be the independent set identified by the greedy algorithm. Let $J = \{q_1, q_2, \ldots, q_r\}$ be a maximum weight independent set. Order both sets by non-increasing weight.

Consider the smallest index k such that $w_{q_k} > w_{e_k}$. If none exists, it follows that $r > m$ in which case we choose $k = m + 1$. In either case we know that $\{q_1, q_2, \ldots, q_k\}$ were not selected by the greedy algorithm in its kth iteration. Furthermore, what was selected had a lower weight. This implies that $q_i \in \{e_1, \ldots, e_{k-1}\}$ or $q_i \cup \{e_1, \ldots, e_{k-1}\} \notin \mathcal{I}$ $\forall i = 1, \ldots, k$. Thus $\{e_1, \ldots, e_{k-1}\}$ is a basis for $\{e_1, \ldots, e_{k-1}, q_1, \ldots, q_k\}$. But $\{q_1, \ldots, q_k\}$ is an independent subset of the same set but of larger size than $\{e_1, \ldots, e_{k-1}\}$. This contradicts Lemma 8.5.   ∎

The theorem is false when $(E, \mathcal{I})$ is not a matroid. Suppose $E = \{1, 2, 3\}$, $w_1 = 1.5, w_2 = 2, w_3 = 1.5$ and $\mathcal{I} = \{\varnothing, (1), (2), (3), (1, 2)\}$. The greedy algorithm returns $\{2\}$, but the maximum weight independent set is $\{1, 3\}$.

8.3   Rank functions

We will be interested in real valued functions defined on subsets of E. These functions will have two properties that will be useful.

Definition 8.11   *Let f be a real valued function on subsets of E.*

- *f is **non-decreasing** if $S \subseteq T \Rightarrow f(S) \leq f(T)$.*
- *f is **submodular** if $\forall S, T \subset E$*

$$f(S) + f(T) \geq f(S \cup T) + f(S \cap T).$$

- *f is **supermodular** if $-f$ is submodular.*

If we assign to each element i of E a real number a_i then $f(S) = \sum_{j \in S} a_j$ and $f(S) = \max_{i \in S} a_i$ are both submodular.

The definition of submodularity given here is identical to the one given in Chapter 7. We associate with each set S a characteristic vector χ_S, i.e., $\chi_S(i) = 1$ if $i \in S$ and zero otherwise. Then submodularity over the characteristic vectors will be:

$$f(\chi_S) + f(\chi_T) \geq f(\chi_S \vee \chi_T) + f(\chi_S \wedge \chi_T).$$

Notice that $\chi_S \vee \chi_T = \chi_{S \cup T}$ and $\chi_S \wedge \chi_T = \chi_{S \cap T}$. The following two results are now trivial.

Theorem 8.12   *f is submodular iff*

$$f(S \cup k) - f(S) \geq f([S \cup k] \cup j) - f(S \cup k)$$

for all $j \neq k$, $j,k \in E$ and $S \subseteq E \setminus \{j,k\}$.

Proof   See the proof of Theorem 7.7.   ■

Theorem 8.12 can be formulated in the following equivalent way:

$$f(S) - f(S \setminus j) \leq f(T) - f(T \setminus j), \quad \forall j \in T \subset S.$$

Theorem 8.12 allows one to interpret submodularity as a discrete form of concavity. If one views $f(S \cup j) - f(S)$ as a derivative, Theorem 8.12 can be interpreted as saying that the derivative decreases as S 'increases'.

Corollary 8.13   *f is submodular and non-decreasing iff*

$$f(T) \leq f(S) + \sum_{j \in T \setminus S} [f(S \cup j) - f(S)], \quad \forall S, T \subseteq E.$$

Here are some properties of submodular functions.

- If f is submodular then $g(S) = f(E \setminus S)$ is submodular.
- If f is submodular and k a number, then $g(S) = \min\{f(S), k\}$ is submodular.
- If f and g are submodular then so is $f + g$.
- If f is submodular then $h(S) = \min\{f(T) : S \subseteq T \subseteq E\}$ is submodular and non-decreasing.

Definition 8.14   *If $(E, \mathcal{I})$ is an independence system, its **rank function**, r, is*

$$r(S) = \max_{T \subseteq S}\{|T| : T \in \mathcal{I}\}.$$

We can describe an independence system by its collection of independent sets, $\mathcal{I}$, or its rank function because $\mathcal{I} = \{T : r(T) = |T|\}$. For this reason we can write an independence system as (E, r).

Theorem 8.15 *$(E, \mathcal{I})$ is a matroid iff its rank function is submodular.*

Proof Let r be the rank function of the matroid $(E, \mathcal{I})$. We show first that r is submodular. Observe first that $r(\varnothing) = 0$, r is non-decreasing and $r(S \cup j) - r(S) \le 1$. To prove submodularity it suffices, by Theorem 8.10, to show that

$$r(S \cup j) - r(S) \ge r(S \cup \{j, k\}) - r(S \cup k).$$

The inequality holds when $r(S \cup j) - r(S) = 1$. So, suppose $r(S \cup j) = r(S) = p$ and $r(S \cup \{j, k\}) - r(S \cup k) = 1$.

Under these assumptions $r(S \cup \{j, k\}) = p + 2, p + 1$. If $r(S \cup \{j, k\}) = p + 2$, then $r(S \cup \{j, k\}) - r(S \cup j) = 2$ which contradicts the fact that $r(S \cup \{j, k\}) - r(S \cup j) \le 1$.

If $r(S \cup \{j, k\}) = p + 1$ then $r(S \cup k) = p$. Let B be a basis for S. Since $r(S \cup j) = r(S \cup k) = p$ it follows that $B \cup j$ and $B \cup k$ are dependent sets. Thus B is a basis for $S \cup \{j, k\}$ which implies that $r(S \cup \{j, k\}) = |B| = r(S) = p$ a contradiction.

Now suppose r, the rank function of the independence system $(E, \mathcal{I})$, is submodular. We show that $(E, \mathcal{I})$ is a matroid. Choose any $S \subseteq E$ and let B and B' be two basis of S with different sizes. Suppose $|B| < |B'|$.

Since r is non-decreasing and submodular

$$r(B') \le r(B) + \sum_{j \in B' \setminus B} [r(B \cup j) - r(B)].$$

Hence $r(B \cup j) > r(B)$ for some $j \in B' \setminus B$. Hence $r(B \cup j) = |B \cup j|$ contradicting the fact that B is a basis. ∎

Definition 8.16 *A real valued function f defined on subsets of E is called a* **matroid rank function** *if*

1. $f(\varnothing) = 0$,
2. *f is integer valued and non-decreasing,*
3. *f is submodular,*
4. *and $f(j) \le 1$ for all $j \in E$.*

Theorem 8.17 *If r is the rank function of the matroid $(E, \mathcal{I})$ then*

$$r^D(S) = |S| + r(E \setminus S) - r(E)$$

is a matroid rank function.

Proof   Clearly $r^D(\varnothing) = 0$, and r^D is submodular. Also

$$r^D(S \cup j) - r^D(S) = 1 - [r(E \setminus S) - r(E \setminus \{S \cup j\})].$$

Hence $0 \leq r^D(S \cup j) - r^D(S) \leq 1$. ∎

Definition 8.18   *If (E,r) is a matroid, the matroid, (E, r^D) is called the matroid* **dual** *to (E,r). In particular B is a basis of (E, r^D) iff $E \setminus B$ is a basis in (E,r).*

8.4   Deletion and contraction

Given a matroid one can obtain two other matroids through two operations called **deletion** and **contraction**.

Definition 8.19   *Let $M = (E, \mathcal{I})$ be a matroid and $S \subseteq E$. The matroid obtained by* **deleting** *S has the ground set $E \setminus S$ and independent set's $\mathcal{I} \setminus S = \{T \subset E \setminus S \colon T \in \mathcal{I}\}$ is denoted $M \setminus S$.*

If r_M is the rank function of M and $r_{M \setminus S}$ the rank function of $M \setminus S$ then $r_{M \setminus S}(T) = r_M(T)$ for each $T \subseteq E \setminus S$.

Definition 8.20   *Let $M = (E, \mathcal{I})$ be a matroid and $S \subseteq E$. Let B be any basis of S. Define M_S, read M* **contract** *S, to be the independence system with ground set $E \setminus S$ and independence family $\mathcal{I}_S = \{T \subset E \setminus S \colon T \cup B \in \mathcal{I}\}$.*

If r_{M_S} is the rank function of M_S then $r_{M_S}(T) = r_M(T \cup S) - r_M(S)$ for all $T \subseteq E \setminus S$.

Theorem 8.21   *M_S is a matroid and its definition does not depend on the choice of B.*

Proof   First we prove that M_S is a matroid. Choose any $A \subseteq E \setminus S$. Let J^1 and J^2 be bases for A with respect to M_S. Therefore $J^1 \cup B$ and $J^2 \cup B$ are independent sets in $\mathcal{I}$. If $J^i \cup B$ is a basis for $A \cup S$ with respect to $\mathcal{I}$ for all i, then we are done since $|J^1 \cup B| = |J^2 \cup B| \Rightarrow |J^1| = |J^2|$.

Suppose not. Then there is an $e \in A \cup S$ with $e \notin J^1 \cup J^2 \cup B$ and $e \cup J^i \cup B$ is in $\mathcal{I}$ for all i. However $e \notin S$ since B is a basis for S with respect to $\mathcal{I}$. Also, $e \notin A$ since J^i is a basis for A with respect to $\mathcal{I}_S$. Therefore $e \notin A \cup S$, a contradiction.

Now we show that M_S does not depend on the choice of B. If it does, S must contain two bases, B^1 and B^2 with respect to M and a set $J \subseteq E \setminus S$ such that $J \cup B^1 \in \mathcal{I}$ but $J \cup B^2$ is not.

Let $J' \subseteq J$ be the largest set such that $J' \cup B^2 \in \mathcal{I}$. Then

$$|J' \cup B^2| < |J \cup B^2| = |J \cup B^1|$$

because $|B^1| = |B^2|$. Thus $J' \cup B^2 \in \mathcal{I}$ is not a basis with respect to M of $J \cup S$. Hence there exists $e \in [J \cup S] \setminus [J' \cup B^2]$ such that $J' \cup B^2 \cup e \in \mathcal{I}$. Now follow the previous argument to show that $e \notin S \cup J$ to derive a contradiction. ∎

Example 33 *Let M be the matroid defined on the columns of the following matrix:*

$$\begin{bmatrix} 1 & 1 & 1 & 0 & 0 & 0 \\ -1 & 0 & 0 & 1 & 0 & -1 \\ 0 & -1 & 0 & -1 & 1 & 0 \\ 0 & 0 & -1 & 0 & -1 & 1 \end{bmatrix}.$$

Choose S to be columns 2 and 4. Now perform the row operations necessary to convert the submatrix associated with columns 2 and 4 into echelon form:

$$\begin{bmatrix} 1 & 1 & 1 & 0 & 0 & 0 \\ -1 & 0 & 0 & 1 & 0 & 1 \\ 0 & 0 & 1 & 0 & 1 & -1 \\ 0 & 0 & -1 & 0 & -1 & 1 \end{bmatrix}.$$

The matroid obtained by contracting S is the matroid defined on the following submatrix:

$$\begin{bmatrix} 0 & 1 & 1 & -1 \\ 0 & -1 & -1 & 1 \end{bmatrix}.$$

8.5 Matroid intersection and partitioning

Here we prove two important theorems about matroids.

Theorem 8.22 (Matroid intersection theorem) *Let $M_1 = (E, \mathcal{I}_1)$ and $M_2 = (E, \mathcal{I}_2)$ be two matroids defined on the same ground set with rank functions r_1 and r_2, respectively. Then*

$$\max\{|J|: J \in \mathcal{I}_1 \cap \mathcal{I}_2\} = \min_{T \subseteq E} r_1(T) + r_2(E \setminus T).$$

Remark Let $J \in \mathcal{I}_1 \cap \mathcal{I}_2$. Then $|J| = |J \cap T| + |J \cap \{E \setminus T\}| \leq r_1(T) + r_2(E \setminus T)$ for all $T \subseteq E$.

Proof   The proof is by induction on $|E|$. The theorem is trivially true for $|E| = 1$. Assume $|E| \geq 2$ and let $k = \min_{T \subseteq E} r_1(T) + r_2(E \setminus T)$. Let $S^1, S^2 = E \setminus S^1$ be a partition of E such that $r_1(S^1) + r_2(S^2) = k$. The proof is divided into two cases.

Case 1:   Exactly one of S^1 or S^2 is empty.

Without loss of generality suppose $S^1 = E$ and $S^2 = \emptyset$. Let $\hat{M}_i$ denote the matroid obtained from M_i by deleting element $j \in E$. By the induction hypothesis there is a set $\hat{J}$ independent in $\hat{M}_i$ for all $i = 1, 2$ and partition of $E \setminus j$ into K^1 and K^2 such that

$$|\hat{J}| = r_1(K^1) + r_2(K^2) \leq r_1(E \setminus j) \leq r_1(E)$$
$$\leq \min\{r_1(K^1 \cup j) + r_2(K^2), r_1(K^1) + r_2(K^2 \cup j)\}.$$

If the right-hand side of the above is $r_1(K^1) + r_2(K^2)$ then we have

$$|\hat{J}| = r_1(K^1) + r_2(K^2) = r_1(K^1 \cup j) + r_2(K^2).$$

Since $\hat{J} \in \mathcal{I}_1 \cap \mathcal{I}_2$, the theorem is proved. If

$$\min\{r_1(K^1 \cup j) + r_2(K^2), r_1(K^1) + r_2(K^2 \cup j)\} = r_1(K^1) + r_2(K^2) + 1$$

it follows that j increases the rank of K^i with respect to r_i for all $i = 1, 2$. Since $K^i \cap \hat{J} \subseteq \hat{J}$ for all i it follows that $r_i(\hat{J} \cup i) = r_i(\hat{J}) + 1$ for all i. Hence $\hat{J} \cup j \in \mathcal{I}_1 \cap \mathcal{I}_2$. Thus

$$|\hat{J} \cup j| = r_1(K^1 \cup j) + r_2(K^2)$$

proving the theorem.

Case 2: $S^1 \neq \emptyset \neq S^2$.

Let $k = r_1(S^1) + r_2(S^2)$. Define M_1' to be M_1 contract S^1 and r_1' the corresponding rank functions. Let $M_2' = M_2 \setminus S^1$ and r_2' the corresponding rank function. For any partition T^1, T^2 of S^2 we have

$$r_1'(T^1) + r_2'(T^2) = r_1(T^1 \cup S^1) - r_1(S^1) + r_2(T^2) \geq k - r_1(S^1) = r_2(S^2).$$

Since $S^1 \neq \emptyset$, $|S^2| < |E|$. Therefore the induction hypothesis applies to S^2. There is a $J^2 \subseteq S^2$ independent in M_1 contract S^1 and $M_2 \setminus S^1$ with $|J^2| = r_2(S^2)$.

A similar argument yields a set $J^1 \subseteq S^1$ independent in $M_1 \setminus S^2$ and M^2 contract S^2 such that $|J^1| = r_1(S^1)$. Notice that $J^1 \cup J^2 \in \mathcal{I}^1 \cap \mathcal{I}^2$ and $|J^1 \cup J^2| = k$ and this proves the claim.   ∎

The matroid intersection theorem does not hold for the intersection of 3 or more matroids. Now we show how the matroid intersection theorem can be used to derive the Hall marriage theorem.

8.5.1   *Application: Hall marriage theorem*

Let A be a set of agents and G a set of goods with $|A| = |G|$. For each agent $i \in A$ there is a non-empty set $D_i \subseteq G$ of goods acceptable to agent i. Given $\{D_i\}_{i \in A}$ is there a way to give every agent $i \in A$ exactly one good in D_i such that no good is assigned to more than agent? Call such an assignment **feasible**. Clearly, not always. This is impossible, for example, when each D_i contains the same single element. Hall's theorem gives a necessary and sufficient condition for a feasible assignment to exist.

If there is a $B \subseteq A$ such that $|B| > |\cup_{i \in B} D_i|$, then there is clearly no feasible assignment. Thus $|B| \le |\cup_{i \in B} D_i|$ for all $B \subseteq A$ is a necessary condition for a feasible assignment. Remarkably it is also a sufficient condition.

Theorem 8.23   *A feasible assignment exists iff $|B| \le |\cup_{i \in B} D_i|$ for all $B \subseteq A$.*

Proof   We prove sufficiency. To avoid trivial cases we assume $|D_i| \ge 2 \ \forall i \in A$ and that for each $j \in G$ there exists i, i' such that $j \in D_i \cap D_{i'}$.

We construct two matroids, M^a and M^g. The common ground set of both matroids is $E = \{(i, j): i \in A, \ j \in D_i\}$. A set $T \subseteq E$ will be independent in M^a iff no agent appears in more than one pair of T. The same set T is independent in M^g iff no good appears in more than one pair of T. Let r^a be the rank function associated with M^a and r^g the rank function associated with M^g.

If T is independent in both matroids, this corresponds to a subset of agents and goods who are paired to each other in one to one fashion. If T is independent in both matroids and $|T| = n$ then T corresponds to a feasible assignment. It suffices to show that there is a set of size n that is independent in both matroids. To derive this from the matroid intersection theorem, it is enough to show that $\min_{T \subseteq E} [r^a(T) + r^g(E \setminus T)] \ge n$ for all $T \subseteq E$.

In an abuse of notation we write $T \cap A$ to denote the set of agents who appear in at least one of the agent–good pairs of T. Similarly with $T \cap G$. Now $r^a(T) = |A \cap T|$ is the number of distinct agents that appear in T. Also $r^g(T) = |A \cap G|$ is the number of distinct goods that appear in T.

Let $T \subseteq E$. Notice that $r^a(T) = |T \cap A|$ and

$$r^g(E \setminus T) = |\{j \in G: \text{s.t. } (i, j) \notin T\}|$$

$$\ge |\{j: j \in D_i, i \notin T \cap A\}|$$

$$= \left| \bigcup_{i \notin T \cap A} D_i \right|$$

$$\ge |A| - |T \cap A|.$$

Hence

$$r^a(T) + r^g(E \setminus T) \geq |A| = n.$$

■

Definition 8.24   *Let $M_i = (E, \mathcal{I}_i)$, $i = 1, \ldots, k$ be a collection of matroids. A set $J \subseteq E$ is **partitionable** with respect to $\{M_i\}_{i \geq 1}$ if there is a partition of J into sets $\{J^1, J^2, \ldots, J^k\}$ such that $J^i \in \mathcal{I}_i$ for all i.*

Theorem 8.25 (Matroid partition theorem)   *Let $M_i = (E, \mathcal{I}_i)$, $i = 1, \ldots, k$ be a collection of matroids with corresponding rank functions r_i. Then*

$$\max\{|J|: J \text{ partitionable}\} = \min_{T \subseteq E} \left\{ |E \setminus T| + \sum_{i=1}^{k} r_i(T) \right\}.$$

Proof   First we prove that

$$\max\{|J|: J \text{ partitionable}\} \leq \min_{T \subseteq E} \left\{ |E \setminus T| + \sum_{i=1}^{k} r_i(T) \right\}.$$

Choose any $J \subseteq E$ and suppose it to be partitioned into sets $F^1, \ldots, F^k$ independent in $M_1, \ldots, M_k$. Choose any $T \subseteq E$. Then

$$|J| = |J \cap (E \setminus T)| + |J \cap T| \leq |E \setminus T| + \sum_{i=1}^{k} |F^i \cap T|$$

$$\leq |E \setminus T| + \sum_{i=1}^{k} r_i(T).$$

To conclude the proof we use the matroid intersection theorem. First we construct two matroids. The first matroid, M', is constructed by making k copies of E. If $E = \{e_1, e_2, \ldots, e_n\}$, then let $E^i = \{e_{i1}, e_{12}, \ldots, e_{in}\}$ be the ith copy of E. The ground set of M' will be $E' = \cup_{i=1}^{k} E^i$. A set $J \subseteq E'$ will be independent in M' iff $J \cap E^i \in \mathcal{I}_i$ for all i. Let r' be the rank function associated with M'. Thus $r'(J) = \sum_{i=1}^{k} r_i(J \cap E^i)$.

To describe the second matroid, let $A^j = \{e_{1j}, e_{2j}, \ldots, e_{kj}\}$ for $j = 1, \ldots, n$. The second matroid, $\hat{M}$, has a ground set E' and $J \subseteq E'$ is independent in $\hat{M}$ iff $|J \cap A^j| \leq 1$ for all j. Let $\hat{r}$ be the rank function of this matroid, then $\hat{r}(J) = |\{j: |J \cap A^j| \geq 1\}|$.

From the matroid intersection theorem we deduce the existence of two sets $J', T' \subseteq E'$ such that

$$r'(T') + \hat{r}(E' \setminus T') = |J'|.$$

where J' is independent in both M' and $\hat{M}$.

From T' we form a new set $\hat{T}$ by removing some elements. If $j \in E$ is such that $A^j \not\subseteq T'$ but $A^j \cap T' \neq \varnothing$ delete $A^j \cap T'$ from T'. For such a j, $1 \leq |E' \backslash T' \cap A^j| \leq |(E' \backslash \hat{T}) \cap A^j|$ and so

$$\hat{r}(E' \backslash T'') = \hat{r}(E' \backslash T').$$

Let $T = \{j \in E: A^j \subseteq \hat{T}\}$ and $J = \{j \in E: A^j \cap J' \neq \varnothing\}$. Since J' is independent in both M' and M'', it follows that $|J'| = |J|$. Then,

$$|E \backslash T| + \sum_{i=1}^{k} r_i(T) \leq \hat{r}(E' - \hat{T}) + r'(T') \leq |J'| = |J|. \qquad \blacksquare$$

One consequence of the partition theorem is that E is partitionable iff

$$\sum_{i=1}^{k} r_i(T) \geq |T|, \quad \forall T \subseteq E.$$

This can be viewed as a generalization of Hall's theorem.

What happens if we ask for a partition of E into bases and not merely independent sets? If $\sum_{i=1}^{k} r_i(E) > |E|$ or $\sum_{i=1}^{k} r_i(E) < |E|$ it cannot be done.

Definition 8.26 *Let $M_i = (E, \mathcal{I}_i)$, $i = 1, \ldots, k$ be a collection of matroids. The collection $\{M_i\}_{i \geq 1}$ can be **packed** into E if there exist disjoint sets $B_1, B_2, \ldots, B_k$ such that B_i is a basis in M_i for all i and $\cup_{i=1}^{k} B_i \subseteq E$.*

Theorem 8.27 (Matroid packing) *Let $M_i = (E, \mathcal{I}_i)$, $i = 1, \ldots, k$ be a collection of matroids. The collection $\{M_i\}_{i \geq 1}$ can be packed into E iff*

$$|T| \leq |E| - \sum_{i=1}^{k} r_i(E) + \sum_{i=1}^{k} r_i(T), \quad \forall T \subseteq E.$$

Proof Suppose first that the collection $\{M_i\}_{i \geq 1}$ can be packed into E. Then there exist disjoint sets $B_1, B_2, \ldots, B_k$ such that B_i is a basis in M_i for all i and $\cup_{i=1}^{k} B_i \subseteq E$. Let $B^0 = E \backslash \cup_{i=1}^{k} B_i$. Choose any $T \subseteq E$. Then

$$|T| = |T \cap B^0| + \sum_{i=1}^{k} |T \cap B_i| \leq |B^0| + \sum_{i=1}^{k} r^i(T)$$

$$= \left| E \backslash \bigcup_{i=1}^{k} B_i \right| + \sum_{i=1}^{k} r^i(T) = |E| - \sum_{i=1}^{k} r_i(E) + \sum_{i=1}^{k} r_i(T).$$

Now suppose that

$$|T| \leq |E| - \sum_{i=1}^{k} r_i(E) + \sum_{i=1}^{k} r_i(T) \quad \forall T \subseteq E. \tag{8.1}$$

We show that the collection of matroids $\{M_i\}_{i=1}^{k}$ can be packed into E.

If we choose $T = \varnothing$ in inequality (8.1) then $t = |E| - \sum_{i=1}^{k} r_i(E) \geq 0$. Define M_{k+1} to be a matroid on E with independent sets $\mathcal{I}_{k+1} = \{S \subseteq E : |S| \leq t\}$. From the matroid partition theorem, E can be partitioned into independent sets from $\{M_i\}_{i=1}^{k+1}$ iff

$$\sum_{i=1}^{k+1} r_i(T) \geq |T|, \quad \forall T \subseteq E.$$

Now

$$\sum_{i=1}^{k+1} r_i(T) = \sum_{i=1}^{k} r_i(T) + r_{k+1}(T)$$

$$= \sum_{i=1}^{k} r_i(T) + \min \left\{ |T|, |E| - \sum_{i=1}^{k} r_i(E) \right\} \geq |T|.$$

Assume such a partition, $\{B_1, B_2, \ldots, B_{k+1}\}$. Each B_i must be a basis in M_i. If not, $|B_i| < r_i(E)$ for one or more i. In this case

$$|E| = \sum_{i=1}^{k+1} |B_i| = \sum_{i=1}^{k+1} r_i(B_i) < \sum_{i=1}^{k+1} r_i(E) = \sum_{i=1}^{k} r_i(E) + t = |E|$$

a contradiction. Hence $\{B_1, B_2, \ldots, B_k\}$ is the packing we seek. ∎

8.6 Polymatroids

Here we show how to formulate the problem of finding a maximum weight basis of an independence system as a linear program. As a first step we formulate the problem as an integer program.

Let $(E, \mathcal{I})$ be the underlying independence system with rank function r. Let $x_j = 1$ if we select $j \in E$ and 0 otherwise. Then:

$$\max \sum_{j \in E} c_j x_j$$

$$\text{s.t.} \quad \sum_{j \in S} x_j \leq r(S), \quad \forall S \subseteq E,$$

$$x_j \in \{0, 1\}, \quad \forall j \in E.$$

Next we relax the condition that $x \in \{0, 1\}$ for all j to $0 \le x_j \le 1$ for all j to produce a linear program (called a linear relaxation of the above):

$$\max \sum_{j \in E} c_j x_j$$

$$\text{s.t. } \sum_{j \in S} x_j \le r(S), \quad \forall S \subseteq E,$$

$$0 \le x_j \le 1, \qquad \forall j \in E.$$

When r is the rank function of a matroid, the extreme points of the feasible region of this linear program are all integral.

Definition 8.28   *Let f be a submodular function on E. The polytope:*

$$P(f) = \left\{ x \in \mathbb{R}_+^n : \sum_{j \in S} x_j \le f(S), \forall S \subseteq E \right\}$$

*is the **polymatroid** associated with (E, f).*

Notice that $P(f) \neq \varnothing$ iff $f(S) \ge 0 \; \forall S \subseteq E$. If, in the definition of $P(f)$, we drop the non-negativity restriction on x, we obtain an **extended polymatroid**, $P^*(f)$.

Theorem 8.29   *Let f be a non-decreasing, integer valued, submodular function on E with $f(\varnothing) = 0$. Then all extreme points of the polymatroid $P(f)$ are integral.*

Proof   We know that for every extreme point, z of $P(f)$, there is a weight vector c such that $cz > cx$ for all $x \in P(f)$. It suffices then to prove that for any choice of weight vector c there is an optimal solution to $\max\{cx : x \in P(f)\}$ that is integral.

Choose a weight vector c and order the elements of E by decreasing weight:

$$c_1 \ge c_2 \cdots c_k \ge 0 > c_{k+1} \cdots \ge c_n.$$

Here we have assumed that k is the largest index for which the corresponding weight is non-negative. Let $S^0 = \varnothing$ and $S^j = \{1, 2, \ldots, j\}$ for all $j \in E$. We show that the vector x defined by $x_j^* = f(S^j) - f(S^{j-1})$ for $1 \le j \le k$ and $x_j^* = 0$ for $j \ge k + 1$ is an optimal solution to the problem $\max\{cx : x \in P(f)\}$.

The vector x^* is clearly integral. It is non-negative because f is non-decreasing. To show that $x^* \in P(f)$ we use the submodularity of f. For any $T \subseteq E$:

$$\sum_{j \in T} x_j = \sum_{j \in T \cup S^k} [f(S^j) - f(S^{j-1})]$$

$$\leq \sum_{j \in T \cup S^k} [f(S^j \cap T) - f(S^{j-1} \cap T)]$$

$$\leq f(S^k \cap T) - f(\varnothing) \leq f(T).$$

The dual to $\max\{cx : x \in P(f)\}$ is:

$$\min \sum_{S \subseteq E} f(S) y_S$$

$$\text{s.t.} \sum_{S \ni j} y_S \geq c_j, \quad \forall j \in E,$$

$$y_S \geq 0, \quad \forall S \subseteq E.$$

To show that x^* is a primal optimal solution it suffices to construct a dual feasible solution with an objective function value of cx^*. To this end set $y_{S^j} = c_j - c_{j+1}$ for $1 \leq j < k$. Set $y_{S^k} = c_k$ and $y_{S^j} = 0$ for $j \geq k+1$.
    Notice that $y_S \geq 0$ for all $S \subseteq E$ and is feasible in the dual because

$$\sum_{S \ni j} y_S = y_{S^j} + \cdots + y_{S^k} = c_j$$

for all $j \leq k$ and $\sum_{S \ni j} y_S \geq 0 \geq c_j$ if $j \geq k+1$.
    The dual objective function value is

$$\sum_{j=1}^{k-1} (c_j - c_{j+1}) f(S^j) + c_k f(S^k) = \sum_{j=1}^{k} c_j [f(S^j) - f(S^{j-1})] = cx^*. \quad \blacksquare$$

If we drop the requirement that f is non-decreasing a similar proof establishes that $P^*(f)$ has all integral extreme points.
    The matroid intersection theorem is also capable of a polymatroidal interpretation. If r_1 and r_2 are the rank functions of two matroids defined on a common

ground set E, then the integer solutions to the following:

$$\sum_{j \in S} x_j \leq r_1(S), \quad \forall S \subseteq E,$$

$$\sum_{j \in S} x_j \leq r_2(S), \quad \forall S \subseteq E,$$

$$0 \leq x_j \leq 1, \qquad \forall j \in E$$

defines a set that is independent in both matroids. An appropriate generalization of the matroid intersection theorem (which we do not discuss here) implies that this polyhedron is integral.

Theorem 8.30   *Let f and g be two integer valued submodular functions defined on the same ground set. Then $P(f) \cap P(g)$ and $P^*(f) \cap P^*(g)$ are integral.*

A separation theorem is also possible.

Theorem 8.31   *Let f and g be two integer valued functions defined on the same ground set E. Suppose f is submodular and g is supermodular such that $f(\emptyset) = 0 = g(\emptyset)$ and $g(S) \leq f(S)$ for all $S \subseteq E$. Then, there exists an integral vector, z in $\mathbb{R}^{|E|}$ such that*

$$g(S) \leq \sum_{j \in S} z_j \leq f(S), \quad \forall S \subseteq E.$$

Proof   We give a proof for the special case when $f(S)$ and $g'(S) = |S| - g(S)$ are both matroid rank functions. To prove the theorem it suffices to show that there is an integral solution to the following system:

$$\sum_{j \in S} x_j \leq f(S), \qquad\qquad \forall S \subseteq E,$$

$$\sum_{j \in S} (1 - x_j) \leq |S| - g(S), \quad \forall S \subseteq E,$$

$$0 \leq x_j \leq 1, \qquad\qquad \forall j \in E.$$

First, we show that if E is partitionable with respect to the matroid M_f associated with f, and $M_{g'}$ the matroid associated with g' then there is an integral solution to the above system. Subsequently we prove that E is partitionable.

If E is partitionable, then we can set $E = L \cup R$ where L is an independent set in M_f and R is an independent set in $M_{g'}$. Now set $z_j = 1$ iff $j \in L$ and zero otherwise. Then for all $S \subseteq E$ we have

$$\sum_{j \in S} z_j = \sum_{j \in S \cap L} z_j = |S \cap L| \le f(S).$$

Second,

$$\sum_{j \in S} (1 - z_j) = |S| - |S \cap L| = |S \cap R| \le g'(S).$$

To prove that E is partitionable, we deduce from the matroid partition theorem that the size of the largest partitionable set is

$$\min_{T \subseteq E} \{|E \setminus T| + f(T) + |T| - g(T)\} = \min_{T \subseteq E} \{|E| + f(T) - g(T)\} = |E|$$

since $f(T) \ge g(T)$ for all $T \subseteq E$ and $f(\varnothing) = g(\varnothing) = 0$. ∎

8.7 Application: efficient allocation with indivisibilities

In this section we generalize the model of Section 4.9. In that model there was a set M of m distinct indivisible objects, and a set N of agents. In Section 4.9 each agent was interested in acquiring no more than one good. Here we will relax this condition.[1]

For every $S \subseteq G$, let $v_j(S)$ be the monetary value that agent $j \in N$ assigns to acquiring S. We assume that each agents valuations are non-negative and do not depend on the goods acquired by other agents.

We impose two conditions on agent's value functions. To describe them we introduce additional notation. Given object prices $p \in \mathbb{R}_+^m$, let the collection of subsets of objects that maximize agent j's utility be denoted $D_j(p)$. Therefore,

$$D_j(p) = \left\{ S \subset M : v_j(S) - \sum_{i \in S} p_i \ge v_j(T) - \sum_{i \in T} p_i \quad \forall T \subset M \right\}.$$

This is agent j's **demand correspondence**. The first condition imposed on value functions is that they be non-decreasing, i.e., for all $j \in N$ and all $S \subset B \subset M$, $v_j(S) \le v_j(B)$.

The second is the **substitutes** (S) condition: For all price vectors p, p' such that $p' \ge p$, and all $S \in D_j(p)$, there exists $B \in D_j(p')$ such that $\{i \in S : p_i = p_i'\} \subset B$.

Example 34 *We give an example of a value function that satisfies (S). For each $i \in M$, let $v(i) = a_i$ where a_i is non-negative. Set $v(S) = \max_{i \in S} a_i$. This is the valuation function of the agents in the model of Section 4.9. Given any price vector $p \in \mathbb{R}_+^m$, the demand correspondence for this value function is $\arg \max_{i \in M}(a_i - p_i)$. It is easy to verify that v satisfies (S). Notice also that v is submodular.*

Theorem 8.32   *Let the value function v be non-decreasing and satisfy (S). Then v is submodular.*

Proof   For any $j \in T \subset S \subset M$ we prove that $v(S) - v(S \setminus j) \le v(T) - v(T \setminus j)$. By Theorem 8.12, this implies submodularity.

Let p be a price vector such that $p_i = 0$ for all $i \in S$ and $p_i > v(M)$ for all $i \in M \setminus S$. With this choice of price vector we see that $S \in D(p)$. Choose any $j \in T$ and $\delta \ge 0$ such that

$$v(S) - \delta = v(S \setminus j).$$

Since v is non-decreasing, such a δ exists.

Let q be a price vector such that $q_i = p_i$ for all $i \in M \setminus j$ and $q_j = p_j + \delta = \delta$. We show that $S \in D(q)$. Certainly there is a $K \in D(q)$ such that $S \setminus j \subset K$ and $K \cap M \setminus S = \varnothing$. If $S \setminus j \in D(q)$ but $S \notin D(q)$ it would mean that

$$v(S \setminus j) = V(S \setminus j) - \sum_{i \in S \setminus j} q_j > v(S) - \sum_{i \in S} q_j = v(S) - \delta,$$

which contradicts the definition of δ.

Let $\hat{p}$ be a price vector such that $\hat{p}_i = 0$ for all $i \in T \setminus j$, $\hat{p}_j = \delta$ and $\hat{p}_i > v(M)$ for all $i \in M \setminus T$. Observe first that $\hat{p} \ge q$. Second, $T = \{i \in M : \hat{p}_i = q_i\}$. Therefore, by (S), $T \in D(\hat{p})$. Hence

$$v(T \setminus j) - \sum_{i \in T \setminus j} \hat{p}_i \le v(T) - \sum_{i \in T} \hat{p}_i.$$

Which simplifies to $v(T \setminus j) \le v(T) - \delta$. Since $\delta = v(S) - v(S \setminus j)$ we conclude that $v(S) - v(S \setminus j) \le v(T) - v(T \setminus j)$.   ∎

The substitutes property implies the following useful property.

Definition 8.33   *A value function, v_j, satisfies the **single improvement** property (SI) if for all $S \notin D_j(p)$ there exists $B \subset M$ such that*

$$v_j(B) - \sum_{i \in B} p_i > v_j(S) - \sum_{i \in S} p_i$$

and $|S \setminus B|, |B \setminus S| \le 1$.

One can interpret (SI) as a kind of local optimality condition. Two sets $A, B \subseteq M$ are considered neighbors if $|A \setminus B| \le 1$ and $|B \setminus A| \le 1$. If $A \notin D(p)$, then (SI) says that there is a neighboring set B that delivers a larger net utility. If $A \in D(p)$, then there is no neighboring set B that delivers strictly more net utility.

Theorem 8.34 *Let v be non-decreasing and satisfy (S). Then v satisfies (SI).*

Proof The proof will be by induction on $|M|$. We leave the reader to verify the base case of $|M| = 2$. Now suppose the theorem true for all $|M| \leq m$. Suppose $|M| = m + 1$. For any price vector p and set $K \subseteq M$ let $g(K, p) = v(K) - \sum_{i \in K} p_i$. Given any $R \subseteq M$ let $D(p, R) = \arg\max_{K \subseteq R} g(K, p)$.

Let p be a price vector and $S \subseteq M$ such that $S \notin D(p, M)$. Since $S \notin D(p, M)$ there is a $T \neq S$ such that $g(T, p) > g(S, p)$. If there is a $k \notin S \cup T$ then $S \notin D(p, M \setminus k)$ and the induction hypothesis applies. Hence for all T such that $g(T, p) > g(S, p)$ we may assume that $S \cup T = M$. In particular for any two sets T and T' such that $g(T, p), g(T', p) > g(S, p)$ it must be that $T \setminus S = T' \setminus S$. Furthermore, for all $K \subset S$ we have $g(K, p) \leq g(S, p)$.

Choose any $T^* \in D(p)$. We show that $|T^* \setminus S| \leq 1$. Suppose not and let $e, f \in T^* \setminus S$. Notice that for any T such that $g(T, p) > g(S, p)$ we have $e, f \in T$. Choose $\epsilon \geq 0$ to solve

$$g(T^*) - 2\epsilon = g(S).$$

Let q be a price vector such that $q_i = p_i$ for all $i \in M \setminus \{e, f\}$ and $q_i = p_i + \epsilon$ for all $i \in \{e, f\}$. Given this choice of ϵ, $g(T, q) = g(T, p) - 2\epsilon$ for all T such that $g(T, p) > g(S, p)$. Hence $S, T^* \in D(q, M)$.

Let q' be a price vector such that $q_i' = q_i$ for all $i \in M \setminus \{f\}$ and $q_f' = q_f + \theta$ where $\theta > \epsilon$ is sufficiently large to ensure that $T^* \notin D(q', M)$. Notice $S \in D(q', M)$. By (S) there is a $K \in D(q', M)$ such that $f \notin K$ and $T \setminus f \subseteq K$. Hence

$$g(S, q') = g(K, q') \Rightarrow g(S, p) = g(K, p) - \epsilon \Rightarrow g(S) < g(K).$$

Therefore $K \setminus S = T^* \setminus S$ a contradiction. Hence $|T^* \setminus S| \leq 1$.

From among all T such that $g(T, p) > g(S, p)$ choose one that maximizes $|S \cap T|$. Call it T'. Note that $|T' \setminus S| = |T^* \setminus S| \leq 1$.

If $|S \setminus T'| \leq 1$ we are done. So, suppose for a contradiction, that $|S \setminus T| \geq 2$. Choose $e \in \arg\min_{i \in S \setminus T'} p_i$ and $\epsilon \geq 0$ to solve

$$g(S, p) = g(T', p) - \epsilon.$$

Now $p_e > 0$. If not, since v is non-decreasing, it follows that $g(T' \cup e, p) \geq g(T', p) > g(S, p)$ contradicting the choice of T'. Let p' be the price vector defined by $p_i' = p_i$ for all $i \notin S \cap T'$ and $p_i' = 0$ for all $i \in S \cap T'$. Observe that $g(S, p') < g(T', p')$ and $g(K, p') \leq g(S, p')$ for all $K \subset S$.

Choose $\delta > 0$ to solve

$$g(S, p^*) + \delta |S \setminus T'| = g(T', p^*).$$

Set $\mu = \min\{\epsilon, \delta\}$. Let q be the price vector defined by $q_i = p_i^*$ for all $i \notin S \setminus T'$ and $q_i = p_i^* - \mu$ for all $i \in S \setminus T'$.

Suppose first that $\epsilon < \delta$. Then

$$g(S, q) = g(S, p^*) + \mu |S \setminus T'| < g(T', p^*) = g(T', q).$$

Since v is non-decreasing and $q_e = 0$ it follows that $g(T' \cup e, q) \geq g(T', q)$. Hence

$$g(T' \cup e, q) > g(S, q) \Rightarrow g(T' \cup e, p^*) > g(S, p^*) \Rightarrow g(T' \cup e, p) > g(S, p)$$

contradicting the choice of T'.

Now suppose that $\epsilon \geq \delta$. Then $g(S, q) = g(T', q)$. Assume first that $S \notin D(q, M)$. Notice that $T' \notin D(q, M)$. Let $K \in D(q, M)$. Since $q_i = 0$ for all $i \in S \cap T'$ we can choose K to contain $S \cap T'$. Now $K \in D(q, M)$ implies

$$v(K) - \sum_{i \in K \cap \{S \cap T'\}} q_i - \sum_{i \in K \cap \{T' \setminus S\}} q_i > v(S) - \sum_{i \in \{S \setminus T'\}} q_i$$

$$\Rightarrow v(K) - \sum_{i \in K \cap \{S \cap T'\}} p_i - \sum_{i \in K \cap \{T' \setminus S\}} p_i > v(S) - \sum_{i \in \{S \setminus T'\}} p_i$$

$$\Rightarrow g(K, p) > g(S, p).$$

Therefore $K \setminus S = T' \setminus S$. However, $K \neq T'$ and so $|K \cap S| > |T' \cap S|$ contradicting the choice of T'.

Now suppose that $S \in D(q, M)$. Increase the price of e only from q_e to p_e. Call the new price vector q^*. Notice $S \notin D(q^*, M)$. By (S) there is a $K \in D(q^*, M)$ such that $S \setminus e \subseteq K$. Hence

$$v(S) - p_e - \sum_{i \in S \setminus e} q_i = g(S, q^*) < g(K, q^*) = g(K, q)$$

$$\Rightarrow v(S) - \sum_{e \in S \setminus T'} p_e < v(K) - \sum_{i \in K \cap \{S \setminus T'\}} p_i - \sum_{i \in K \cap \{T' \setminus S\}} p_i$$

$$\Rightarrow g(S, p) < g(K, p).$$

Since $|K \setminus S| = 1$ this contradicts the choice of T' and the proof is complete. ∎

An assignment of goods to agents such that no good is assigned to more than one agent will be called an **allocation**. An allocation is **efficient** if it maximizes the total realized value. We will show how to formulate the problem of finding an efficient allocation as a linear program. From the duality theorem of linear programming and complementary slackness we will identify, as we did in Section 4.9, supporting prices.

We first formulate the problem problem of finding an efficient allocation as an integer program. Let $y_j(S) = 1$ if agent j is to be allocated the bundle $S \subseteq M$, and zero otherwise. The optimization problem, denoted (**P**), is to solve

$$V(A) = \max \sum_{j \in N} \sum_{S \subseteq M} v_j(S) y_j(S)$$

$$\text{s.t.} \quad \sum_{S \ni i} \sum_{j \in N} y_j(S) \le 1, \quad \forall i \in M,$$

$$\sum_{S \subseteq G} y_j(S) \le 1, \qquad \forall j \in N,$$

$$y_j(S) \in \{0, 1\}, \qquad \forall S \subseteq M, \ \forall j \in N.$$

The first constraint ensures that overlapping sets of goods are never assigned. The second ensures that no agent receives more than one subset.

We will relax the constraint $y_j(S) \in \{0, 1\}$ to $0 \le y_j(S) \le 1$ for all $S \subseteq M$ and $j \in N$. We will show that this linear relaxation has an integer optimal solution.

Given a value function v_j let $K_j(p) = \min\{|A|: A \in D_j(p)\}$ and

$$D_j^*(p) = \{A \in D_j(p): |A| = K_j(p)\}.$$

Lemma 8.35 *Let v_j be non-decreasing and satisfy (S) and let*

$$\phi_j(T, p) = \min_{S \in D_j(p)} |S \cap T|.$$

Then

$$\phi_j(T, p) = \min_{S \in D_j^*(p)} |S \cap T|.$$

Proof Choose any $S \in D_j(p)$. We show that there exists a $B \in D_j^*(p)$ such that $B \subseteq S$. Suppose not, then $|B \setminus S| > 0$. From amongst all such B choose one that minimizes $|B \setminus S|$. Let $r \in B \setminus S$.

From p construct a new price vector p' by raising prices on all elements in $E \setminus \{S \cup B\}$ by some large amount (exceeding the maximum value achievable on any bundle) and raising the price on r by $\epsilon > 0$.

Clearly $S \in D_j(p')$ but $B \notin D_j(p')$. By (SI) there exists $C \subseteq A \cup B$ such that $C \in D_j(p')$ and $C \cup B = B \setminus b$ and $|C \setminus B| \le 1$. In particular $|C| \le |B|$ and $|C \setminus S| = |B \setminus S| - 1$. If $C \in D_j^*(p)$, this would contradict the choice of B.

Now

$$C \in D_j(p') \Rightarrow v_j(C) - \sum_{j \in C} p'_j > v_j(B) - \sum_{j \in B} p'_j.$$

Therefore,

$$v_j(C) - \sum_{j \in C} p_j > v_j(B) - \sum_{j \in B} p_j - \epsilon.$$

Since this inequality holds for all sufficiently small ϵ it follows that $C \in D_j(p)$. Since $|C| \leq |B|$ we conclude that $C \in D_j^*(p)$ yielding the desired contradiction.

∎

This lemma allows us to construct a matroid M_j with M as the ground set. The bases of this matroid will be all sets in $D_j^*(p)$. A set S will be independent iff $S \subseteq B$ for some $B \in D_j^*(p)$. It is easy to check that this forms an independence system. The matroid property is satisfied by definition since all bases have size $K_j(p)$. The rank function r_j, for this matroid is:

$$r_j(S) = \max_{B \in D_j^*(p)} |B \cap S|.$$

However, $\max_{B \in D_j^*(p)} |B \cap S| = K_j(p) - \min_{B \in D_j^*(p)} |B \cap \{M \setminus T\}|$. Since $K_j(p) = \phi_j(M, p)$ and in view of Lemma 8.35 $r_j(S) = \phi_j(M, p) - \phi_j(M \setminus S, p)$.

We show that linear programming relaxation of (P) has an optimal integer solution. Relaxing the integrality constraint in (P) gives

$$V_{LP}(A) = \max \sum_{j \in N} \sum_{S \subseteq M} v_j(S) y_j(S)$$

$$\text{s.t.} \ \sum_{S \ni i} \sum_{j \in N} y_j(S) \leq 1, \quad \forall i \in M,$$

$$\sum_{S \subseteq M} y_j(S) \leq 1, \qquad \forall j \in N,$$

$$y_j(S) \geq 0, \qquad \forall S \subseteq M, \ \forall j \in N.$$

Theorem 8.36 *If each agent's value function is monotonic and satisfies (SI),* $V(A) = V_{LP}(A)$.

Proof Let p^* be an optimal dual solution (here p^* corresponds to the first set of constraints) to the linear relaxation of (P). Let

$$V_p(A) = \max \sum_{j \in N} \sum_{S \subseteq M} v_j(S) y_j(S) - \sum_{i \in M} p_i \left[\sum_{S \ni i} \sum_{j \in N} y_j(S) \right] + \sum_{i \in M} p_i$$

$$\text{s.t.} \ \sum_{S \subseteq M} y_j(S) \leq 1, \quad \forall j \in N,$$

$$y_j(S) \geq 0, \qquad \forall S \subseteq M, \ \forall j \in N.$$

The objective function can be written as

$$V_p(A) = \max \sum_{j \in N} \sum_{S \subseteq M} \left[v_j(S) - \sum_{i \in S} p_i \right] y_j(S) + \sum_{i \in M} p_i.$$

In this problem, if $y_j(S) = 1$ is part of an optimal solution, then $S \in D_j(p)$.

By the duality theorem of linear programming, $V_{p^*}(A) = \min_{p \geq 0} V_p(A) = V_{LP}(A)$. If there is a partition of M so that each agent j receives at most one element of $D_j(p)$, it would follow that (P) has an optimal integer solution.

Suppose no such partition exists. In particular there is no partition of M that gives each agent j an element of $D_j^*(p)$. In matroid language, there is no partition of M into sets $\{B_1, B_2, \ldots, B_{|A|}\}$ such that each B_i is a basis in M_i. From the matroid packing theorem we deduce the existence of a $M \setminus T \subseteq M$ such that

$$|M \setminus T| > |M| - \sum_{i=1}^{|N|} r_i(E) + \sum_{i=1}^{|N|} r_i(M \setminus T),$$

in other words,

$$\sum_{j \in A} \phi_j(T, p^*) > |T|.$$

Define a new dual solution p such that $p_i = p_i^*$ for all $i \in M \setminus T$ and $p_i = p_i^* + \epsilon$ for all $i \in T$.

For each $j \in N$, let $B^j \in D_j(p)$ be such that $|B^j \cap T| = \phi_j(T, p)$. If $\epsilon > 0$ is chosen sufficiently small we claim that $\sum_{j \in A} |B^j \cap T| > |T|$. Suppose for the moment this is true. Then,

$$V_p(A) = \sum_{j \in N} \left[v_j(B^j) - \sum_{i \in B^j} p_i \right] + \sum_{i \in M} p_i^* + \epsilon |T|$$

$$= \sum_{j \in N} \left[v_j(B^j) - \sum_{i \in B^j} p_i^* \right] - \epsilon \sum_{j \in N} |B^j \cap T| + \sum_{i \in M} p_i^* + \epsilon |T|$$

$$\leq \sum_{j \in N} \left[v_j(C^j) - \sum_{i \in C^j} p_i^* \right] - \epsilon \sum_{j \in N} |B^j \cap T| + \sum_{i \in M} p_i^* + \epsilon |T|$$

$$= V_{p^*}(N) - \epsilon(|B^j \cap T| - |T|)$$

$$< V_{p^*}(N) = V_{LP}(N)$$

which contradicts the optimality of p^*.

It remains then to prove the claim. Let $C^j \in D_j(p^*)$ be such that $|C^j \cap T| = \phi_j(T, p^*)$. For any $S \subset E$ such that $|S \cap T| < |C^j \cap T|$ we must have

$$\left[v_j(C^j) - \sum_{i \in C^j} p_i^* \right] - \left[v_j(S) - \sum_{i \in S} p_i^* \right] > 0.$$

Let ϵ^j be the minimum of the left-hand side over all such sets S. Observe that $\epsilon^j > 0$. Now increase the price of all elements in T by $\epsilon^j/|T|$. Call the new price vector p'. We show there is a $B^j \in D_j(p')$ such that $|B^j \cap T| = |C^j \cap T|$. Suppose, for a contradiction that $|B^j \cap T| < |C^j \cap T|$ for all $B^j \in D_j(p')$. Then

$$v_j(B^j) - \sum_{i \in B^j} p_i' = v_j(B^j) - \sum_{i \in B^j} p_i^* - \frac{\epsilon^j}{|T|} |B^j \cap T|$$

$$> v_j(C^j) - \sum_{i \in C^j} p_i^* - \frac{\epsilon^j}{|T|} |C^j \cap T|.$$

Since $v_j(C^j) - \sum_{i \in C^j} p_i^* > v_j(B^j) - \sum_{i \in B^j} p_i^* + \epsilon^j$ we deduce that

$$v_j(B^j) - \sum_{i \in B^j} p_i^* - \frac{\epsilon^j}{|T|} |B^j \cap T| > v_j(B^j) - \sum_{i \in B^j} p_i^* + \epsilon^j - \frac{\epsilon^j}{|T|} |C^j \cap T|,$$

i.e.,

$$\frac{\epsilon^j}{|T|} |C^j \cap T| - \frac{\epsilon^j}{|T|} |B^j \cap T| > \epsilon^j$$

which is a contradiction. To complete the proof of the claim it suffices to repeat this argument for each agent j with $\epsilon = \min_j \epsilon^j$.   ∎

8.8   Application: Shannon switching game

Let $G = (V, E)$ be a graph (possibly with parallel edges) and e^* a distinguished element of E. There are two players, called the **cut** and **short** player. The cut player moves first followed by the short player and thereafter they alternate. On any one of her moves, the cut player can delete any element from E that has not previously been 'tagged' by the short player. On each of his moves, the short player tags an element not previously deleted or tagged. Element e^* can never be deleted or tagged. The goal of the short player is to tag a subset of S of E such that $S \cup \{e^*\}$ contains a cycle through e^*. The goal of the cut player is to prevent this. Clearly, one of the two players must win and there are no ties. The game was invented by Claude Elwood Shannon (1916–2001), of Information Theory fame. The game is related to Hex discussed earlier in this book.

Example 35   *An instance of the game is depicted in Figure 8.1.[2] We show that the short player has a strategy that guarantees a win. Since the cut player moves first, she must delete e_1. If not, the short player will tag e_1 and win the game. The short player tags e_5. The cut player must respond by deleting e_6 to prevent the short player from winning. Next, the short player tags e_2. This forces the cut player to delete e_7. The short player tags e_3 next. The cut player must now delete one of e_4 or e_8. No matter which one the cut player deletes, the short player can tag the other to win the game.*

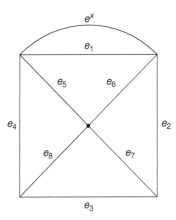

Figure 8.1

Here is a variant of the game played on a vector space rather than graph. Let E be a finite set of vectors and e^* a distinguished element of E. The goal of the short player is to tag a subset S of E such that $e^* \in \text{span}(S)$. The goal of the cut player is to prevent this.

The game can be played on any matroid by suitably defining what it means for an element to be contained in the span of a set.

Definition 8.37   *Let $M = (E, \mathcal{I})$ be a matroid with rank function r. The **span** of $S \subset E$ is*

$$\text{span}(S) = \{j \in E : r(A \cup j) = r(A)\}.$$

If M is a graphic matroid, the span of a set S will be the set of all edges e such that $e \in S$ or $S \cup e$ contains a cycle.

Lemma 8.38   *Let $M_i = (E, \mathcal{I}_i)$, $i = 1, \ldots, k$ be k copies of the same matroid M. Let J be a maximum cardinality partitionable set. Then there are disjoint* *independent sets $F_1, F_2, \ldots, F_k$ such that*

$$E \setminus J \subseteq \mathrm{span}(F_1) = \cdots = \mathrm{span}(F_k).$$

Proof If $J = E$, the lemma follows by setting $F_1 = F_2 = \cdots = F_k = \varnothing$. Since J is a maximum cardinality partitionable set, it follows from the matroid partition theorem that there is a set $T \subseteq E$ such that

$$|J| = |E \setminus T| + kr(T).$$

Let $P_1, P_2, \ldots, P_k$ be the partition of J into independent sets. Then

$$|J| = |J \cap (E \setminus T)| + |J \cap T| \leq |E \setminus T| + \sum_{i=1}^{k} |T \cap P_i|$$

$$\leq |E \setminus T| + kr(T) = |J|.$$

Thus all inequalities must hold at equality. The first inequality binding in this chain implies that $|J \cap \{E \setminus T\}| = |E \setminus T|$, i.e., $E \setminus T \subseteq J$. The second inequality binding in this chain implies that $\sum_{i=1}^{k} |T \cap P_i| = kr(T)$, i.e., $r(T) = |T \cap P_i|$ for all i.

Set $F_i = T \cap P_i$ for all i. These F_i's are independent. Since $r(T) = |F_i|$ for all i it follows that $T \subseteq \mathrm{span}(F_i)$ for all i. As $E \setminus T \subseteq J$ we deduce that $E \setminus J \subseteq \mathrm{span}(F_i)$ for all i. Finally, $F_j \subseteq T \subseteq \mathrm{span}(F_i)$ for all $i \neq j$. Thus $\mathrm{span}(F_1) = \mathrm{span}(F_2) = \cdots = \mathrm{span}(F_k)$. ■

Theorem 8.39 *Let M be a matroid with distinguished element e^*. The Shannon game on (M, e^*) can be won by the short player iff there exist disjoint subsets $F^1, F^2 \subseteq E \setminus e^*$ such that $e^* \in \mathrm{span}(F^1) = \mathrm{span}(F^2)$.*

Proof Suppose first that there exist disjoint subsets $F^1, F^2 \subseteq E \setminus e^*$ such that $e^* \in \mathrm{span}(F^1) = \mathrm{span}(F^2)$. We exhibit a winning strategy for the short player.

Without loss of generality we can assume that $F^1 \cup F^2 = E \setminus e^*$ and that F^1 and F^2 are bases for E. Set $F_0^1 = F^1$ and $F_0^2 = F^2$. Let $e_1 \in F_0^1$ be the first element deleted by the cut player. Since M is a matroid, by Lemma 8.5 there is a $f_1 \in E$ such that $(F_0^1 \setminus e_1) \cup f_1$ is a bases for E. The short player tags f_1 and updates F_0^1, F_0^2 as follows:

$$F_1^1 = (F_0^1 \setminus e_1) \cup f_1, \quad F_1^2 = F_0^2.$$

After k rounds of play, sets F_k^1 and F_k^2 have been defined such that

1. F_k^1, F_k^2 are bases of E,
2. $F_k^1 \cap F_k^2$ are the set of elements tagged after k turns, and
3. $[F_k^1 \cup F_k^2] \setminus [F_k^1 \cap F_k^2]$ is the set of unplayed elements.

By (2) if $e^* \in \text{span}(F_k^1 \cap F_k^2)$, the short player has won. If not, the cut player selects an $e_k \in [F_k^1 \cup F_k^2] \setminus [F_k^1 \cap F_k^2]$. Say, $e_k \in F_k^1$. By Lemma 8.5 there is an $f_k \in F_k^2$ such that $(F_k^1 \setminus e_k) \cup f_k$ is a bases. The short player tags f_k and updates F_k^1, F_k^2 as follows:

$$F_{k+1}^1 = (F_k^1 \setminus e_k) \cup f_k, \quad F_{k+1}^2 = F_k^2.$$

Thus after each round, two elements are consumed and $F_k^1 \cap F_k^2$ is increased by one element.

Since $|E|$ is finite, the process must terminate. If it terminates with $e^* \in \text{span}(F_k^1 \cap F_k^2)$ the short player wins. If not, by (3) all elements have been played, i.e., $F_k^1 = F_k^2$. But from (1) $e^* \in \text{span}(F_k^1) = \text{span}(F_k^1 \cap F_k^2)$ and again the short player wins.

Now suppose the Shannon game can be won by the short player. Let J be a maximum cardinality subset of E partitionable into two M-independent sets. If we can choose J so that $e^* \notin J$, the proof is complete because we have identified disjoint independent sets $F^1, F^2 \subseteq E \setminus e^*$ such that $e^* \in \text{span}(F^1) = \text{span}(F^2)$. If not, we exhibit a winning strategy for the cut player.

Assume first that $J = E$. Let S^1 and S^2 be disjoint independent sets that partition E. Suppose $e^* \in S^1$. Extend S^1 and S^2 into bases B_1 and B_2 with $e^* \notin B_2$. Then $B_i^* = E \setminus B_i$ are disjoint bases in the dual matroid M^D. Notice that $e^* \in B_2^*$. Since M^D is a matroid there is an $f \in B_1^*$ such that $B_2^* \setminus e^* \cup f$ is a bases of M^D. Set $F_1 = B_1^*$ and $F_2 = B_2^* \setminus e^* \cup f$.

Consider now the matroid M^D contract f, i.e., M_f^D. This has ground set $E \setminus f$. Observe that F_1 and F_2 are disjoint bases in M_f^D and that $e \notin F_1 \cup F_2$. Furthermore $e^* \subset \text{span}(F_1) = \text{span}(F_2)$ where the span is relative to M_f^D.

The cut player now plays as follows. On her first move she deletes f. From then on she plays as if she was the short player but playing the game on M_f^D. Thus cutting an element in M corresponds to tagging an element in M_f^D. Given the sets F_1 and F_2 just constructed the cut player has a strategy that will guarantee that she can cut a set T such that e^* is in the span of T with respect to M_f^D. If r_f^D is the rank function of M_f^D it follows that

$$r_f^D(S) = |S| + r(E \setminus S) - r(E),$$

where r is the rank function of M. Since T spans e^* in M_f^D it follows that $r_f^D(T \cup e^*) = r_f^D(T)$, i.e.,

$$|T \cup e^*| + r(E \setminus (T \cup e^*)) - r(E) = |T| + r(E \setminus T) - r(E),$$

i.e., $r(E \setminus T) = r(E \setminus (T \cup e^*)) + 1$. Hence no subset of $E \setminus T$ spans e^* which contradicts the fact that the short player has winning strategy.

Last suppose $J \neq E$ but $e^* \in J$. Since J was chosen to exclude e^* if possible it follows that $J \setminus e^*$ is a maximum cardinality partitionable set of the matroid M delete e^*. By Lemma 8.38 there exist disjoint independent sets, F_1, F_2 (in M delete e^*) such that $e^* \notin \text{span}(F_1) = \text{span}(F_2)$ and $E \setminus (J \cup e^*) \subseteq \text{span}(F_1) = \text{span}(F_2)$.

Let M' be the matroid M contract $\text{span}(F_1)$. If the short player has a winning strategy in M he has a winning strategy in M'. However $E \setminus \text{span}(F_1)$ is partitionable in M' and so by the previous argument the cut player has a winning strategy. ∎

Problems

8.1 Let E be a ground set and k a positive integer. Let $\mathcal{I} = \{S \subseteq E \colon |S| \leq k\}$. Show that $(E, \mathcal{I})$ is a matroid. This is called the **uniform** matroid.

8.2 Let $(E, \mathcal{I})$ be an independent system. Suppose for any $A, B \in \mathcal{I}$ with $|A| < |B|$ exists $j \in B \setminus A$ such that $A \cup j \in \mathcal{I}$. Show that $(E, \mathcal{I})$ is a matroid.

8.3 Let $(E, \mathcal{I})$ be an independent system. Suppose for any two bases of E, A and B the following was true:

For all $i \in A \setminus B$ there is a $j \in B \setminus A$ such that $\{A \setminus i\} \cup j$ and $\{B \setminus j\} \cup i$ are bases of E.

Show that $(E, \mathcal{I})$ is a matroid. Prove the converse as well.

8.4 Let $(E, \mathcal{I})$ be a matroid with weight vector w that assigns to each $e \in E$ a weight w_e. Let

$$f(K) = \max_{S \in \mathcal{I}, S \subseteq K} \sum_{e \in S} w_e.$$

Show that f is submodular.

Notes

1 This section is based on Gul and Stachetti (2000).
2 This example is from Bixby (1982).

References

Bixby, R. E.: 1982, Matroids and operations research, *in* H. J. Greenberg, F. H. Murphy and S. H. Shaw (eds), *Advanced techniques in the practice of operations research*, North Holland, New York.

Gul, F. and Stacchetti, E.: 2000, The english auction with differentiated commodities, *Journal of Economic Theory* **92**(1), 66–95.

Oxley, J. G.: 1992, *Matroid theory*, Oxford Science Publications, Oxford University Press, Oxford, New York.

Index

Milton Keynes UK
Ingram Content Group UK Ltd.
UKHW052047240923
429320UK00002B/2

9 780415 700085